SECOND EDITION

The Vietnam War

For Those Who Still Care
An Engineer's Story

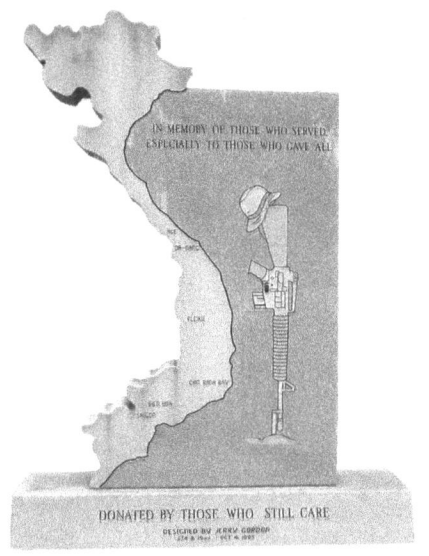

Colonel Ivan V Beggs (Retired, USAR)

Copyright © 2023 by Ivan Beggs with the US Copyright Office.

Copyright #: TXu2-364-838

All rights reserved. No part of this publication may be reproduced, distributed, or transmitted in any form or by any means, including photocopying, recording, or other electronic or mechanical methods, without the prior written permission of the publisher, except in the case of brief quotations embodied in critical reviews and certain other noncommercial uses permitted by copyright law.

The Vietnam War - For Those Who Still Care by Colonel Ivan V Beggs (USAR). Revised title from "For Those Who Still Care".

Editors: Brenda Parker, Asheville, NC; Bradley Yoder, Indianapolis, IN
Book Designer – Creative Publishing Book Design,
   www.creativepublishingdesign.com
Cover creation: by Ivan Beggs, Brenda Parker, and Creative Publishing Book
   Design – All rights reserved by Ivan Beggs

"For Those Who Still Care" monument, Old Court House, Hendersonville, NC designed by Jerry Gordon FEB. 8, 1947 – OCT. 4, 1995[1]

ISBN: 978-1-7341167-6-2 Soft cover

HIS027070 **HISTORY** / Wars & Conflicts / Vietnam War
BIO008000 **BIOGRAPHY & AUTOBIOGRAPHY** / Military
BIO026000 **BIOGRAPHY & AUTOBIOGRAPHY** / Personal Memoirs

Images used in the book are used with permission listed in the annex and usually in the footnote on page where the image is shown.

Second Edition, 2024

14 13 12 11 10 / 10 9 8 7 6 5 4 3 2

---

[1] *https://docsouth.unc.edu/commland/monument/586/*

Contents

Acknowledgements v
Vietnam War Veterans Monument vii
Book Dedication xi
Foreword xii
Introduction. 1
Chapter 1—Problems with the War 3
Chapter 2—Welcome to Vietnam 45
Chapter 3—Intelligence Officer 51
Chapter 4—Company Commander 61
Chapter 5—Moving the Company 69
Chapter 6—The 101st Airmobile Div 77
Chapter 7—Other Activities 97
Chapter 8—Going Home 109
A Political Suggestion 115
Annex – Vietnam Statistics 117
Table of Figures 123
Permissions List 127
About the Author 135

Acknowledgements

Many thanks to Marlene, my wife for fifty-three years, for her patience and especially our children.

Thank you 1LT Ken Ament for writing Chapter 5, "Moving the Company," for leading the movement by LST, and for permissions to use your photographs. And especially to your pouring your heart and soul in doing excellent work with the Company D, 84th Engineer Battalion operations. That paid off afterwards in your civilian life by founding and being President of Construction Control Corporation, Salt Lake City, Utah. It provided premier construction management and cost control. Projects such as all construction at the Salt Palace, Utah's largest public building, and the world renowned Salt Lake City Main Library. Construction Control Corporation has provided cost consulting and estimating services on over 3,500 construction projects since 1984. These projects range up to $200 million in construction costs, and cover a wide variety of building types. Projects such as Higher Education, K-12 education, Civic, Performing Arts, Office, Bio Medical, Healthcare, Laboratories, and Manufacturing. [1]

Many thanks to 1LT Donald Schlotz who was a thoughtful quiet leader and counselor to all of us in Company D, 84th Engineer Battalion. Sadly he passed away before his time. He will be missed.

Thank you 1LT Chuck Stewart for the many conversations from the Battalion Operations staff viewpoint. I and many others have valued your input and appreciate your work keeping the Battalion together for fifty-three years.

Colonel Melvyn Remus – Thank you for your firm gracious mentoring in Vietnam and in the past five decades. You have been very influential!

---

[1] https://cccutah.com/

And many thanks to Brenda Parker who meticulously and painstakingly polished my prose. She is an editor, who saw the big picture of the book, suggested many detailed improvements, and guided me in restructuring and shaping it. For the right writer, she is an outstanding editor.

Vietnam War Veterans Monument
Dedicated to all who had
Served
Died
& were Wounded

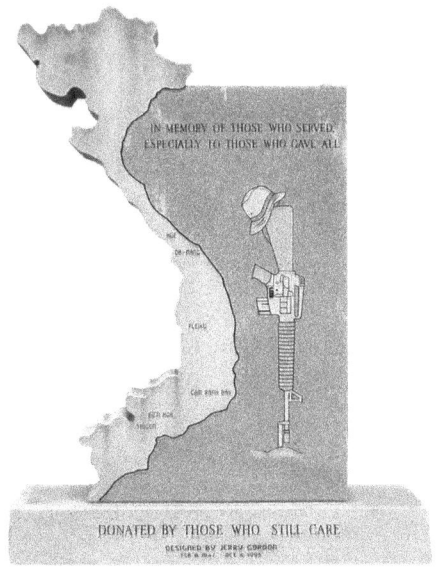

Located at
Henderson County Heritage Museum
The Old Henderson County Court House
1 Historic Courthouse Square
Henderson County, NC

Thus, this book is titled

"For Those Who Still Care"

Henderson County Vietnam Memorial, Hendersonville, NC[1]

The front of the two-toned gray granite monument shows a rifle embedded upright in the ground with a forage cap hanging on the gun's stock. The left side of the monument in the lighter gray is carved in the shape of North and South Vietnam. The back of the marker has the names of 16 Henderson County men who died in the war. This marker is one of nine memorials honoring Henderson County citizens that form "The Honor Walk" on the grounds of the old Henderson County Courthouse.

IN MEMORY OF THOSE WHO SERVED, / ESPECIALLY TO THOSE WHO GAVE ALL / DONATED BY THOSE WHO STILL CARE / DESIGNED BY JERRY GORDON / FEB. 8, 1947 – OCT. 4, 1995

The original dedication was held in the mid-1990's, but on Sunday afternoon, April 13, 2008, this memorial and eight other markers honoring Henderson County war dead were dedicated or rededicated. This ceremony was the culmination of a three day event to celebrate the ten million dollar restoration of the historic Henderson County Courthouse and its rededication as the home of the Henderson County Heritage Museum.

The Hendersonville Community Band played patriotic tunes while the crowd of over 200 sang along, prayed and listened to speakers and watched veterans lay wreaths on monuments honoring the fallen in all wars in American history. "What mean these stones?" asked George A. Jones, chairman of the Henderson County Heritage Museum Board. "We have erected them in honor of all of these. This should never be made in a lighthearted or frivolous manner. There's too much blood, too much sacrifice, too much death they represent," he answered to his own question. The Disabled American Veteran Veterans, Shytle-Beddingfield Chapter placed a wreath on this monument during the ceremony.

---

[1] https://docsouth.unc.edu/commland/monument/586/

The monument is located by the Henderson County Heritage Museum (Old Henderson County Courthouse) at 1 Historic Courthouse Square, Hendersonville NC. Several monuments are nearby, including Revolutionary Soldiers Memorial, Confederate Soldier Memorial, Union Soldiers Monument, War Memorial, World War I, World War II, Korean War, Gulf War, Iraq and Afghanistan monuments.

Backside of the Monument

On the back of the monument are the names of 16 men from Henderson County, North Carolina, who died in the US Vietnam War:

JOHN W. IRVING JR	MAY 8, 1965
JAMES PIERPOINT MORGAN	MAR 11, 1966
GEORGE CLYDE BEDDINGFIELD	DEC. 9, 1966
ARTHUR WAYNE WILKIE	JAN. 16, 1967
BOBBY DEAN LIVELY	AUG 7, 1967
LARRY GAY LANCE	SEPT. 8, 1967
KENNETH RAY LANCASTER MIA	JAN. 3, 1968
BILLY LEWIS STRICKLAND	JAN. 24, 1968
CHARLES LEE KING	APR. 4, 1968
JAMES DANIEL BAILEY	JAN. 17, 1969
ROY EDWARD PITTS	FEB. 17, 1969
RICHARD LEE WAYCASTER	FEB. 19, 1969
GARY RICHARD MASON	SEPT. 14, 1969
ROBERT LAMAR JOHNSON	JAN. 25, 1970
PAUL ROBERT STEPP JR.	MAR. 29, 1970
RICHARD BRYAN LINEBERRY	SEPT. 27, 1972

They are part of the:

  58,156 who also died

  303,704 wounded in action

  9,087,000 military personnel served on active duty during the Vietnam Era (28 February 1961 - 7 May 1975)

  82% of veterans who saw heavy combat strongly believe the war was lost because of a lack of political will

  Nearly 75% of the general public (in 1993) agrees with that.[1]

---

[1] https://www.vva310.org/vietnam-war-statistics

Book Dedication

This book is especially dedicated to all those who served in Vietnam, particularly those who were injured, to the 58,000 Americans who made the ultimate sacrifice, and to the quiet families back home. And to the South Vietnames soldiers, their families, and civilians who fought for their freedom.

Also, it is dedicated to those that I was directly involved with in Vietnam:

- Colonel Melvyn Remus
- Colonel Harry MacGregor
- First Lieutenant Ken Ament
- First Lieutenant Donald Schlotz (deceased)
- First Lieutenant Charles Stewart
- First Lieutenant 'Hatch' Hatcher
- Chief Maintenance Officer whose name was "Chief"
- First Sergeant whose name was "Top"

At one point I knew the names of almost all the soldiers I worked with. Unfortunately, fifty-three years later their names but not their faces have faded from memory. Sorry. Thank you for your work, sacrifice, arguments, and discussions. We did do our jobs and we went home.

This book is also dedicated to these people although we did not serve together in Vietnam:

- Major General James J. Hughes Jr
- Colonel Henry J. Thayer (deceased)
- Colonel Edward Guthrie (deceased)
- Chaplain (Colonel) E.H. Jim Ammeran (deceased)
- Chaplain (Colonel) James E. Wright (deceased)
- Captain Richard Hillier
- Captain Roger T. Heimann
- First Lieutenant Bob Holyfield (deceased)
- CSM Paul Walby (deceased)
- Bradley L.Yoder, Professor of Sociology & Social Work, Manchester College.

Foreword

As the war dragged on, anyone who could do so avoided serving. As is always the case, those whose parents had money or influence had strings pulled for them and stayed safely at home. Many others burned their draft cards and fled to Canada. These young men are now grandfathers and the war that they were a part of has been over now for 50 years. Most of them put the war behind them and went on to live their lives, some never recovered from the trauma; but for all of them, the war was an experience that shaped their lives, even if no one wanted to hear about their experiences in it. They have stories that they never got to tell at the time and experiences for which they rarely, if ever, received appropriate recognition but which deserve to be heard so that no one forgets what they endured.

This is a story of a young 24 year old captain who was responsible for the lives of 180 young men in a war zone. Sensibly making the most of a rotten situation, his main objective was to make sure that he and his men did the jobs they'd been assigned and be able to come home safely. They were engineers and not fighters, so he considers that he and his men were lucky.

Ivan served in our nation's most contentious war. I protested against it, but I'm honored to have been entrusted with helping to organize the memories and recollections of a soldier who did his duty despite his doubts about the validity of his mission. I may not agree with everything that he's written, but this is his story and not mine, and my responsibility as his editor was to guide and suggest, not to lecture and preach. It's my hope that by reading his words maybe people will get a better understanding of what the war was about and why it was so controversial and provoked such extreme reactions throughout both the U.S. and most of the western world.

And, Ivan, for what it's worth, some of us do still care.

Brenda Parker

Introduction

Prior to entering the Army, I was barely aware of the swirling political storms in the US and the world such as the Cold War between the Western Powers and the Soviet Union, the Vietnam War, racial issues, social unrest, and antiwar protests. Instead, I was highly focused on surviving engineering high school and engineering college as well as jogging, bicycling, swimming, religion, and girls.

So, I entered the US Army in September 1968 for training at Fort Belvoir, Virginia, as an engineer officer. Ten weeks later in Germany I met my future wife of fifty-three years. She was an American school teacher to the children of the US Armed Forces. Five months after we were married I received orders to arrive in Vietnam on September 10, 1970.

Eventually I was assigned as an engineer company commander. None of us in my company were dramatic heroes nor had experiences like those shown in movies such as, "Apocalypse Now," "Platoon," "Saving Private Ryan," "The Longest Day," the various Rambo movies, or other such movies.

Instead, we were engineers quietly building or repairing roads, bridges, airfields, culverts, buildings, or whatever else that was needed by the direct combat units such as the 101st Airmobile Division. They were the ones who had the excitement, terror, boredom, and the recognition of real war that is popularized in movies.

So, this is my story, which is really a part of their story. Few of us understood the context of the Vietnam War. We were there because the US government sent us there. However, neither the President, the Congress, religious leaders, business leaders, and generals effectively told us the background of the war, nor why we were in Vietnam. In our own ways we trusted them. So, we just did our jobs and went home.

Thus, Chapter 1 explains what I and many of us did not understand and what the major problems of the war were. The remaining chapters tell some of my and our experiences from September 10, 1970 until September 9, 1971.

Chapter 1
Problems with the War

What amazes me is that in 1970-1971 most of the soldiers, like me, were only vaguely aware of the history of the Western powers in Vietnam, the Domino Theory for the War, US government lies, race riots, anti-war protests, Jane Fonda, the Gulf of Tonkin Incident, Agent Orange, and the Mai Lai Massacre.

Consequently, soldiers repeatedly asked me as the Company Commander, "Why are we here?"

My response, which I intensely did not like, was, "Let's do our jobs and go home." And that is what we did. Through disagreements and arguments, we did our jobs, went home and nobody cared.

It wasn't until decades later when I had the time and the interest to understand the War, that these issues became clear to me.

This chapter explains the context of the war that I, and probably the soldiers in my unit, did not understand. My and our experiences follow in Chapters 2-8.

The French in Vietnam

In the later part of the 1800's and the earlier part of the 1900's Great Britain, France, Belgium, Holland, Spain, Portugal, Germany, Japan, and to some extent the United States, competed to build empires around the world. The French claimed Vietnam as a colony in 1877.

Consequently, the Vietnamese effort for independence was nearly a one hundred year effort. Prior, to becoming President of the Democratic State of Vietnam at the end of the Second World War, Ho Chi Minh had been one of the outspoken voices for Vietnamese independence while in France during World War I.[1]

During World War II, France temporarily lost control of the area when the Japanese invaded. After the War, the French attempted to reassert control. The Vietnamese leader Ho Chi Minh declared independence on September 2, 1945, shortly after the Japanese surrender. The Indochina War between Vietnam and France lasted from 1945 to 1954. With substantial support from Chairman Mao in China, Ho Chi Minh defeated the French at the Battle of Dien Bien Phu in May 1954.

At the end of World War II, there was a world conflict between capitalism and communism. The communist countries, particularly the USSR, fervently wanted the world to become communist and to liberate the world from the chains of capitalism. The Western powers, lead by the United States, wanted the world to be free of the tyranny of totalitarian communism. Consequently, both sides spent massive amounts of money on their militaries. As a result, there were sporadic small wars scattered around the world such as the Chinese Communist Revolution (1945-1949), the Berlin Blockade (1948-1949), the Korean War (1950-1953), the Hungarian Revolution (1956), the Suez Canal Crisis (1956), the Berlin Crisis (1961), the Cuban Missile Crisis (1962), and then the US-Vietnam War (1964-1975). During the same time period, the US and the Soviet Union competed politically in Latin America, the Middle East, Africa, Asia, and Oceania.[2]

The Viet Minh forces were formally created in May 19, 1941 by the Indochinese Communist Party to form the Democratic State of Vietnam. They exhorted "soldiers, workers, peasants, intellectuals, civil servants, merchants, young men and women to overthrow the French jackals and the Japanese fascists." The Việt Minh established itself as the only organized anti-French and anti-Japanese resistance group. It eventually seized the northern Vietnamese city of Hanoi and declared a Democratic State of Vietnam (or North Vietnam) with Ho Chi Minh as president for the next 25 years.[3] But, the French were reluctant to leave and so, they fought Ho Chi Minh's forces.

---

[1] https://www.history.com/topics/vietnam-war/ho-chi-minh-1, accessed February 18, 2023.

[2] https://en.wikipedia.org/wiki/Cold_War

[3] https://en.wikipedia.org/wiki/Viet_Minh

However, as President of the United States, the very highly respected General of the Army, Dwight Eisenhower neither supported nor interfered with the French involvement in South Vietnam nor helped to rescue the French in 1954.[4] He also did not advocate for involvement in Vietnam to prevent it from being over run by the Viet Minh communist forces, which later became the Viet Cong.

However, he helped the United States to assume responsibility from France for protecting South Vietnam, negotiated the Southeast Asia Treaty Organization (SEATO), conditionally pledging the United State to protect Indochina, and gave $7 billion in economic and military aid to South Vietnam from 1955-1961.

In April 1954 Eisenhower stated, "The possible consequences of the loss (of Indochina) are just incalculable to the free world....that if Indochina falls, the rest of Southeast Asia would "go over very quickly" like a "row of dominoes...."[5]

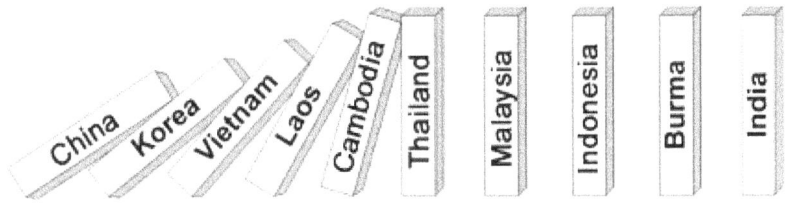

Figure 1-1 Domino Theory – When the first one falls, they all fall.

The Domino Theory states that when a set of dominos is stood up in a line or curve, tipping over one domino will knock the next one down, which knocks the next one down, until there are no more to knock down.

However, it meant little to the soldiers that I knew. It did not motivate the US population who, like me, did not grasp the emotional, political, ideological, and social catastrophe of a communist victory in Southeast Asia, let alone of the Cold War. Some politicians and the major military leaders talked about the Domino Theory. That if Vietnam falls to the Communists, then all of Southeast Asia will fall to the Communists. Then it won't be long before the rest of the world is Communist and then the US will also fall.

---

[4] McNamara, Robert S., *In Retrospect – The Tragedy and Lessons of Vietnam,"* page 358.

[5] McNamara, In Retrospect, page 31

Surprisingly, Laos was more important than Vietnam at the time. A RAND (abbreviation for Research & Development) Corporation study summarized the Laos as: "Hardly a nation except in the legal sense. It lacked the ability to defend its recent independence. Its economy was undeveloped, its administrative capacity primitive, its population divided both ethnically and regionally, and its elite disunited, corrupt, and unfit to lead." But this weak state was the "cork in the bottle," as President Dwight D. Eisenhower, the outgoing US President, summarized in his meeting with President John F. Kennedy, the incoming President, "Its loss to be "the beginning of the loss of most of the Far East."[6]

Previously, the Eisenhower administration had worked for years to create a strong anti-Communist bastion in Laos, a bulwark against Communist China and North Vietnam. While attractive on a map, this strategy was completely at odds with the characteristics of the Laotian state and people.[7]

The transition meetings between the outgoing Eisenhower administration and the incoming Kennedy administration gave a mixed conclusion to the Vietnam situation. Thus, Eisenhower said, "If Laos were lost, all of Southeast Asia would fall. By implication, the West would have to do whatever was necessary to prevent that outcome."

Therefore, President Kennedy in his 1961 Inaugural Address said, "Let every nation know, whether it wishes us well or ill, that we shall pay any price, bear any burden, meet any hardship, support any friend, oppose any foe to assure the survival and the success of liberty. This much we pledge–and more."[8]

By that time China and North Korea had fallen to Communism. The fear was that if South Vietnam also fell to Communism, then the rest of Southeast Asia (Laos, Cambodia, Thailand, etc. as shown in the diagram) would also fall. Therefore, the Domino Theory became the underlying strategy for the US involvement in the war. So, at the end of 1960 there were 685 military

---

[6] https://history.state.gov/milestones/1961-1968/laos-crisis
[7] https://history.state.gov/milestones/1961-1968/laos-crisis
[8] https://ushistoryscene.com/article/president-john-f-kennedys-inaugural-address-1961/
https://www.ourdocuments.gov/doc.php/print_friendly.php?flash=false&page=&doc=91&title=President+John+F.+Kennedys+Inaugural+Address+%281961%29

advisors in South Vietnam[9]. Later President Kennedy boosted the number of advisors to 20,000.

None of the US Presidents, nor US politicians, nor the US government effectively sold the War to the American public let alone to the world public. There were no propaganda type films, no effective political campaigns, nothing. There was just a vague concept called the Domino Theory.

Consequently, who wanted to be severely injured or die for the Domino Theory? Who was willing to send their children to fight, or die, or be injured for the Domino Theory? While people understood and willingly supported standing up to the Communist threat in Europe, there was little home support to send 'our boys' to a strange land on the other side of the world that most of us had never even heard of. When it was a few advisors who were volunteers they did not care nor pay much attention. However, after some years of the draft, television showing daily films of the war, the deaths, the destruction, and the lack of political leadership, most support for the war dwindled and then evaporated.

President Johnson tried to end the War. In 1964 he said, "We are not about to send American boys nine or ten thousand miles away from home to do what Asian boys ought to be doing for themselves."[10]

Three years later he said, "Our purpose in Vietnam is to prevent the success of aggression. It is not conquest, it is not empire, it is not foreign bases, it is not domination. It is, simply put, just to prevent the forceful conquest of South Vietnam by North Vietnam."[11]

By the time I was in Vietnam from September 10, 1970 to September 9, 1971, these statements did not motivate the soldiers, their families, nor the nation to support the War. Some of the soldiers I knew felt, "Who wants to

---

[9] https://www.gale.com/binaries/content/assets/gale-us-en/primary-sources/archives-unbound/primary-sources_archives-unbound_u.s.-military-advisory-effort-in-vietnam_military-assistance-advisory-group-vietnam-1950-1964.pdf

[10] Speech at Akron University, Akron, Ohio, 21 Oct. 1964 https://www.azquotes.com/author/7511-Lyndon_B_Johnson/tag/vietnam-war

[11] Johnson, Lyndon B. (1967). "Public Papers of the Presidents of the United States: Lyndon B. Johnson, 1966", p.211, Best Books on https://www.azquotes.com/quote/148323

die for a war that our country is not supporting while the South Vietnamese are corrupt and have an incompetent military?"

My response, right or wrong was, "Let's do our jobs, be safe, and go home." Through all our emotional trials, arguments, mistakes and successes, that is what we did. We worked together, went home, and nobody cared.

Nixon – "Peace in Our Time"

The war was still going on when I left South Vietnam but in an address to the nation on April 26, 1972, which became known as "Peace in Our Time," President Nixon modified the aim of the U.S. forces. He said, "My fellow Americans, let us therefore unite as a nation in a firm and wise policy of real peace–not the peace of surrender, but peace with honor–not just peace in our time, but peace for generations to come."[12]

My guess is that this is what later motivated General Colin Powell, Chairman of the Joint Chiefs of Staff, to formulate the Powell Doctrine, to engage in a war "…regarding American national security interests, with overwhelming force, and widespread public support."[13]

What the Presidents during the Vietnam War era didn't do was to get the public support for the war, nor do it with overwhelming force. Thus, this ineffective political strategy supported Chairman Mao's strategy of Protracted War which was to wear out an opponent who does not have the willingness to fight a hundred years war. Furthermore, Mao advocated attacking where the enemy is weak. Withdraw from where they're strong. Avoid direct conflict with a strong force.

Additionally, Ho Chi Minh said, "You can kill ten of our men for every one we kill of yours. But even at those odds, you will lose, and we will win."[14] Ho Chi Minh understood that there was a lack of political support for the War in the US, while he had strong support in North Vietnam and even some support in South Vietnam.

---

[12] https://www.nixonfoundation.org/2022/04/peace-generations-come/ Accessed Dec 4, 2022.
[13] https://en.wikipedia.org/wiki/Colin_Powell
[14] https://www.brainyquote.com/quotes/ho_chi_minh_347067

It did not help the American efforts that while the South Vietnamese had some very capable units, most of their military needed many more years before they were fully trained, capable, and effectively led. Since the American public increasingly shared similar views and no President campaigned to win the support of the American public the chances of the US winning the War were doomed.

This was a pervasive feeling with the soldiers I knew. For most of them, it was a vague sense. For a few soldiers it was strong feeling. Nevertheless, the we continued to do our jobs.

Secretary of Defense McNamara

The US Secretary of Defense, Robert McNamara, made several crucial statements in his book, *"In Retrospect – The Tragedy and Lessons of Vietnam."* That if the South Vietnamese were to succeed they had to have the will to win the war. That external forces such as the United States cannot be a substitute for political order and stability. Instead the people themselves must fix their political issues.[15] The general feeling was that the regime was corrupt and inept. That the primary problem in South Vietnam was political and military.[16] For instance, in 1964 there were three governments in three months.[17] The populace did not have faith in their leaders.

"Furthermore, McNamara stated, "I criticize the President, his advisers, and myself as much as the Chiefs for this negligence. It was our job to demand the answers. We did not press hard enough for them. And the Chiefs did not volunteer them. General Bruce Palmer, Jr, Vice Chief of Staff of the US Army, said that, 'Not once during the war did the Joint Chiefs of Staff advise the Commander-In-Chief (the President of the United States) or the Secretary of Defense that the strategy would probably fail and that the US would be unable to achieve its objectives."[18]

McNamara continued, "A major part for that failure was having many more commitments than just the situation in Vietnam. Instability in Latin America,

---

[15] Page 333, McNamara, *In Retrospect, 1996.*
[16] Page 108, McNamara, *In Retrospect, 1996*
[17] Pages 112-113, McNamara, *In Retrospect, 1996*
[18] Page 108, McNamara, *In Retrospect, 1996,*

Africa, and the Middle East, and the Soviet threat in Europe all took up time and attention. There was no senior group exclusively assigned to resolve the situation in Vietnam; so, the crisis there became just one of many items on each person's plate. When combined with the inflexibility of our objectives, and the fact that we had not truly investigated what was essentially at stake and important to our interests, we were left harried, and overburdened.' We never stopped to fully explore whether there are other options."

More significantly, "The Joint Chiefs stated that we fought on the enemy's terms and obliged ourselves with self-imposed restrictions. Nevertheless, the broadening of the war with US air attacks on North Vietnam and shifting from training the South Vietnamese to carrying out the war in both South and North Vietnam with U.S. combat forces, was done with two and half pages that had little analysis or supporting rationale".[19]

"What was also unfortunate, is that the US did not have any specialists of South Vietnam in neither the Department of Defense nor the U.S. State Department. Thus, there was no one who had in depth knowledge that could guide the senior level government of the United States."[20]

Also in 1966, the Joint Chiefs war-gamed the situation. They concluded that the Viet Cong[21] strategy of avoiding major engagements with U.S. forces would make it extremely difficult to find and fix enemy units….Viet Cong experience in the jungles and with guerrilla warfare would pose serious problems, even for well-equipped and highly mobile U.S. regulars. They believed that the punishment being imposed could and would be absorbed by the Hanoi leadership because the country was a subsistence economy centering on self-sustaining villages.[22] And that the air attacks would not work.[23]

---

[19] Pages 108-109, McNamara, *In Restrospect, 1996*
[20] Page 117, McNamara, *In Restrospect, 1996*
[21] The Viet Cong -- also known as the National Liberation Front – was primarily an army organization based in Southern Vietnam and parts of Cambodia. The Viet Cong was formed in 1954 and lasted until 1976, when it was disbanded. While the organization also had political leadership, it was primarily concerned with military action and comprised the forces that fought against the United States and its allies during the Vietnam War. https://classroom.synonym.com/difference-between-viet-cong-and-viet-minh-12083891.html
[22] Page 208 McNamara, *In Restrospect, 1996*
[23] Page 114, McNamara, *In Restrospect, 1996*

As the Washington Post wrote in a review of his book, *In Retrospect – The Tragedy and Lessons of Vietnam*, "He (McNamara) and his colleagues, including Secretary of State Dean Rusk and national security adviser McGeorge Bundy, were not stupid or venal. Dubbed "the best and the brightest," they were all smart, dedicated people who "acted according to what we thought were the principles and traditions of this nation. Yet we were wrong, terribly wrong. We owe it to future generations to explain why." His answer mostly is that they could not figure out what to do, so they just blundered ahead, sustained by wishful thinking."[24]

The Pentagon Papers

"On June 1967, Secretary of Defense Robert McNamara commissioned a sweeping study of the Vietnam War that would later become known as "The Pentagon Papers." It created a major political firestorm."

"McNamara had been a leading proponent of US involvement in Vietnam, but by 1967, he was disillusioned with the war and no longer believed in the policies he had been so instrumental in establishing."

"The study filled 47 volumes, a total of 7,000 pages. Of these, 3,000 pages were historical studies and the other 4,000 pages were government documents. The official title was "US-Vietnam Relations, 1945-1967: History of US Decision Making Process on Vietnam Policy." It was dubbed "The Pentagon Papers" by the news media in 1971."

"Only 15 copies of the study were produced.[25] A leaked copy created a major political firestorm. More details are in the referenced footnote.

Gulf of Tonkin

In 1964, under the guidance of the US Department of Defense, the South Vietnamese Navy attacked communications stations, bridges, and other targets. In support of these operations, a destroyer, the USS Maddox, monitored communications in the area. Several North Vietnamese torpedo

---

[24] https://www.washingtonpost.com/archive/politics/1995/04/09/mcnamara-writes-vietnam-mea-culpa/a85cc058-54fe-4074-bda3-b374885ede8f/
[25] https://www.airforcemag.com/article/0207pentagon/

boats were also in the area. The Maddox was fired upon, escaped damage, and a torpedo boat was damaged. The North Vietnamese view was that the Maddox was conducting operations and was a legitimate target.

However, there were conflicting reports about whether the attack had or had not occurred. Many observers felt that President Johnson was looking for a reason to increase the war effort and was looking for an incident to arouse public and Congressional support.[26]

There was a feeling that the incident did not rise to the level of justifying a full-scale war. On the other hand, others felt that it was imperative to stop the communist aggression before it became too great. Thus, some believed that the government lied to get the US into the war, while others believed that the US was justified in order to stop the spread of communism.

On August 5, 1964, Congress passed The Gulf of Tonkin Resolution authorizing the President to take all necessary actions to repel any attacks against US forces, prevent further aggressions, and maintain peace and security in Southeast Asia, which was vital to US interests and world peace.[27]

Agent Orange

In guerrilla warfare, the guerrillas are not a stand-alone army. For every guerrilla unit there are people, farms, businesses, villages, and cities that provide weapons, food, clothing, medical supplies, ammunition, and intelligence.

In mid-1961 President of South Vietnam, Ngo Dihn Diem, requested the US to conduct herbicide operations. In November, President Kennedy authorized the program Operation Ranch Hand.[28]

Between 1962 and 1971, the United Stated sprayed nearly 20,000,000 gallons of a variety of herbicides in Vietnam, eastern Laos, and Cambodia, 2/3 of which were 13,000,000 gallons of Agent Orange[29] called Operation Ranch Hand.

---

[26] https://www.britannica.com/event/Gulf-of-Tonkin-incident
[27] https://www.britannica.com/event/Gulf-of-Tonkin-Resolution
[28] https://en.wikipedia.org/wiki/Agent_Orange#cite_note-27
[29] https://www.history.com/topics/vietnam-war/agent-orange-1

The objective was to deprive the guerrillas getting food from rice paddies and farmlands, and to destroy concealment along roads, canals, and rivers from ambushes, as well as concealment of their bases, and to clear the brush around US bases.[30]

Figure 1-2 Army Huey helicopter spraying Agent Orange over agricultural land

The US Army used mostly helicopters and low flying aircraft as well as armored personnel carriers, trucks, and backpacks to spray Agent Orange and other such herbicides.[31] Operation Ranch Hand sprayed more than 20% of the forests. The main ingredient Dioxin persists in soil, lake and river sediments and in the food chain. It also accumulates in the fatty tissue of fish, birds, other animals. Thus, humans consuming meat, poultry, dairy, eggs, shellfish, and fish as well coming into contact by spray or from the sprayed vegetation absorb Dioxin.[32]

In 1988, Dr. James Clary, an Air Force researcher associated with Operation Ranch Hand, wrote to Senator Tom Daschle, "When we initiated the herbicide program in the 1960s, we were aware of the potential for damage due to Dioxin contamination in the herbicide. However, because the material was to be used on the enemy, none of us were overly concerned. We never considered a scenario in which our own personnel would become contaminated with the herbicide."[33]

The US soldiers were told that the chemicals being sprayed were defoliates to deprive the guerillas of a food supply and should not cause any issues.

However, several of the soldiers in my unit said to me, "Sir, we are farm boys. We use chemicals to control pests and weeds. This isn't bug spray to kill mosquitoes. What is this stuff?"

---

[30] https://en.wikipedia.org/wiki/Agent_Orange#cite_note-27
[31] https://commons.wikimedia.org/wiki/File:US-Huey-helicopter-spraying-Agent-Orange-in-Vietnam.jpg https://www.vietnam.ttu.edu/virtualarchive/items.php?item=VA042084
[32] https://www.history.com/topics/vietnam-war/agent-orange-1
[33] https://www.history.com/topics/vietnam-war/agent-orange-1

Trusting the soldiers, I investigated. I found an airbase where these flights came from. People there told me several times, "Sir, this is a bug spray to kill the mosquitoes."

Several decades later several of my friends from different units in Vietnam suffered the consequences of Agent Orange. Teeth falling out. Facial features decaying. Heart failure. After much coaxing or harassing on my part, one eventually signed up with the Veterans Administration (VA). They sent him to the Cleveland Medical Center where he had a complete heart transplant and gained a new life. Instead of silently decaying like a weed, he was able to recapture some of his life back. There were several other such soldiers that I encouraged to sign up with the VA with similar experiences though not as severe as just mentioned.

In my unit, we were fortunate. Nearly every night we had access to showers and relatively clean clothes. We were able to get the road dust and grime from our engineering operations off of us. Unknowingly we were fortunate while others were not. Yet, decades later, I don't know if anyone in the Company suffered from Agent Orange.

Decades later, an Air Force Master Sergeant got very upset with me for not signing up with the VA. So, I reluctantly toddled over to the nearest office. The person taking my information was also upset with me for not signing up earlier. A few weeks later, I had a physical, and received in the mail a thick packet. A letter said, "If you have any of these diseases, it is presumptively assumed that they are from Agent Orange exposure in Vietnam. The VA will pay for all costs associated with the diseases." I casually thumbed through the half inch document and shrugged my shoulders, sarcastically thinking, "Right…."

However, my wife, bless her heart, carefully read the whole document. I thought she was going to pass out. She was visibly upset and concerned about my health; more so, than I was. I am very lucky to have such an attentive spouse. Not everyone does. That is probably why I have survived so long. The letter mentioned something else. "Your children may also suffer from your exposure to Agent Orange…."That caused me concern which continues to this day.

Here is a partial list of some of the diseases associated with Agent Orange and to which U.S. G.I.s and Vietnamese soldiers and civilians were exposed to as of 2020:

> Chronic B-Cell Leukemia
> Hodgkin's disease
> Multiple Myeloma
> Non-Hodgkin's lymphoma
> Prostate cancer
> Respiratory Cancers
> Soft tissue sarcomas
> Ischemic heart disease
> Chloracne
> Porphyria cutanea tarda
> Parkinson's disease
> Peripheral neuropathy
> Type 2 Diabetes Mellitus
> AL Amyloidosis[34]

See the Appendix for more enlightening and disheartening details.

The US disputes the South Vietnamese government's estimates that three million Vietnamese have been adversely affected by Agent Orange and related chemicals and the US disputes the Red Cross's estimate that one million have suffered.[35] Furthermore, Vietnam claims half a million children have been born with serious birth defects.[36]

Figure 1-3 A person with birth deformities associated with prenatal exposure to Agent Orange

The picture shows a person with birth deformities associated with prenatal exposure to Agent Orange. He is begging for money while he shows his severe arm deformity most likely

---

[34] https://www.hillandponton.com/agent-orange-and-your-body-symptoms/
[35] https://en.wikipedia.org/wiki/Agent_Orange#cite_note-27
[36] https://www.history.com/topics/vietnam-war/agent-orange-1

related to agent orange exposure when he was in gestation and his pregnant mother was exposed to the defoliating chemical, dioxin.[37]

Pictured here is Ho Chi Minh Professor Nguyen Thi Ngoc Phuong, at Tu Du Obstetrics and Gynecology Hospital. She is pictured with a group of handicapped children, most of them victims of Agent Orange.[38]

Figure 1-4 Handicapped children, most of them victims of Agent Orange

Neither the soldiers nor I were aware of the information presented here. There was some suspicion that there was a problem, but we basically trusted the system. Yet…there were doubts.

Mai Lai Massacre

In 1968 the North Vietnamese made an all out assault across all of South Vietnam. As part of the counter offensive a platoon was ordered on March 16, 1968 to eradicate anyone who was a communist in the village of Mai Lai.[39] The platoon killed between 347 to 504 men, women, children, and infants. Some were gang raped. Twenty-six soldiers were charged, with only Lieutenant William Calley Jr found guilty of killing twenty-two villagers. He was given a life sentence but served three-and-a-half years under house arrest. The chain of command attempted to cover up the incident. When the news broke it contributed to more of the anti-war movement.[40] The event stimulated many to call GI's 'baby killers.'

Figure 1-5 The aftermath of the Mai Lai massacre showing mostly women and children dead on a road, March 16, 1968

---

[37] Emilio Labrador from Davie (South Florida), USA - Agent Orange Deformities. https://commons.wikimedia.org/wiki/File:Agent_Orange_Deformities_(3786919757).jpg
[38] https://commons.wikimedia.org/wiki/File:A_vietnamese_Professor_is_pictured_with_a_group_of_handicapped_children.jpg
[39] https://www.youtube.com/watch?v=n4Qr8oW1QS4
[40] https://en.wikipedia.org/wiki/M%E1%BB%B9_Lai_massacre
https://en.wikipedia.org/wiki/Hugh_Thompson_Jr.

Chapter 1: Problems with the War   17

Figure 1-6 The photo became a symbol of antiwar movement.

While all this was reported in the media, I was more concerned about graduating from college, working with the US Army in Europe, enjoying fun trips exploring Europe, and courting my future wife. Also, while I was stationed in Germany, the only English news media I had access to was the independent newspaper Stars and Stripes, and Armed Forces Radio. Thus, I was only vaguely aware of Mai Lai.

Antiwar Protests

"On June 11, 1963, a Buddhist monk Thích Quảng Đức burned himself to death at a busy Saigon road intersection in protest against Diệm's policies.[41] In response to Buddhist self-immolation as a form of protest, Madame Nhu—the *de facto* First Lady of South Vietnam at the time (and the wife of Ngô Đình Nhu, who was the brother and chief advisor to Diệm)—said

Figure 1-7 Typical student protesters marching at the University of Wisconsin-Madison during the Vietnam War era (1965)

"Let them burn and we shall clap our hands," and "If the Buddhists wish to have another barbecue, I will be glad to supply the gasoline and a match."" [42]

As the war progressed, the antiwar protests increased. So, the soldiers I knew in Vietnam were aware of the unpopularity of the war back home in the US.

As the news media showed the war each night on televisions in

---

[41] https://en.wikipedia.org/wiki/Th%C3%ADch_Qu%E1%BA%A3ng_%C4%90%E1%BB%A9c#/media/File:Th%C3%ADch_Qu%E1%BA%A3ng_%C4%90%E1%BB%A9c_self-immolation.jpg
[42] https://en.wikipedia.org/wiki/Buddhist_crisis

American homes and the draft became stronger, the protests gathered more participants. Many were on campuses across the US. An early one was of students marching down Langdon Street at the University of Wisconsin-Madison during the Vietnam War era (1965)[43]

Figure 1-8 King speaking to an anti-Vietnam war rally at the University of Minnesota in St. Paul, April 27, 1967

Figure 1-9 Mohammad Ali versus George Foreman, October 30, 1974, "Rumble in the Jungle."

Martin Luther King was a Baptist pastor to the African American community and the prominent personality campaigning for civil rights for his community. His people did not have equal rights for voting, housing, financing, education, jobs, and daily normal interactions with many of the white community. He felt, Why should the Negro fight in a distant war when he cannot have those rights in his homeland. Eventually, he felt that the war was unjust. Consequently, while campaigning for civil rights for his people. King began to increasingly speak out against the war.[44]

A famous heavyweight world champion boxer, Mohammad Ali,[45] also protested that as a African American man he couldn't see why he should go and fight in a white man's war when African American

---

[43] Student_Vietnam_War_protesters.JPG(510 × 450 pixels, file size: 71 KB, MIME type: image/jpeg) https://commons.wikimedia.org/wiki/File:Student_Vietnam_War_protesters.JPG#/media/File:Student_Vietnam_War_protesters.JPG

[44] https://en.wikipedia.org/wiki/Beyond_Vietnam:_A_Time_to_Break_Silence#/media/File:Martin_Luther_King_Jr_St_Paul_Campus_U_MN.jpg
CC BY-SA 2.0 File:Martin Luther King Jr St Paul Campus U MN.jpg Created: 27 April 1967

[45] https://upload.wikimedia.org/wikipedia/commons/5/5f/Ali_hitting_foreman

people aren't free in the US. He said, *"I ain't got no quarrel with them, Viet Cong."* The media vilified him, the government prosecuted him for draft dodging, and the boxing commission took away his boxing license.[46]

As the war continued the protests against it grew larger and larger. On January 20th, 1969 a protest released balloons each one representing a US service man killed in Vietnam.[47]

Figure 1-10 Symbolic representation of American dead in Vietnam since Jan. 20th 1969

The student protests became more emotional and violent. However, usually not reported and overlooked is the increasing violence by protesters. Four months before I arrived in Vietnam, students at Kent State University, Kent, Ohio, for three days attacked and set fire to various buildings including the Reserve Officer Training Corp building. Some also attacked the National Guardsmen that were sent to quell the riots. A few used long needles to stick Guardsmen. The violence increased the tensions among the Guardsmen.[48]

A picture of Mary Ann Veccio gesturing and screaming as she kneels by the body of a student, Jeffrey Miller, lying face down on the campus of Kent State University, in Kent, Ohio on May 4, 1970. It is a historically significant photo of the event and one of the defining images of the Vietnam War. It

One of the most famous pictures of the protest movement is of Mary Ann Veccio screaming by the body of Jeffry Miller. It won a Pulitzer Prize.

I do not have permission from the photographer John Paul Filo to use the picture. View it at:

https://en.wikipedia.org/wiki/File:Kent_State_massacre.jpg

Figure 1-11 Guardsmen killed a rioter during the riots at Kent State University May 4, 1970

---

[46] https://aaregistry.org/story/muhammad-ali-stripped-of-title-for-opposing-vietnam-war/
[47] https://upload.wikimedia.org/wikipedia/commons/0/0d/%22Lie_down_and_be_counted%22_Anti-Vietnam_War_Demonstration.jpg
By Hartmut Schmidt, Heidelberg - Own work, CC BY-SA 4.0, https://commons.wikimedia.org/w/index.php?curid=93086152
[48] Mentioned to me by a few soldiers who were then on the KSU Campus.

became a Pulitzer prize-winning photo. There exist multiple, non-trivial mentions of the image in notable, mainstream sources. For example, CNN describes it as a "famous photograph", and NPR describes it as an "iconic photo."[49]

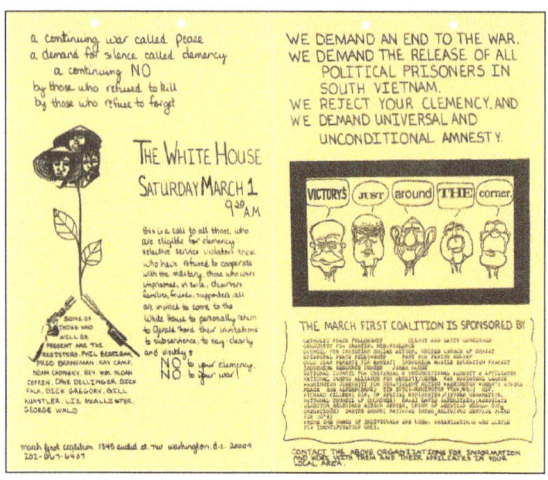

Figure 1-12 Flyer for antiwar march, Washington, D.C., March 1, 1975. "March First Coalition

Announcing protests was done by word of mouth, ads in papers, in churches, and handwritten flyers. A famous antiwar march was conducted years after I left Vietnam in Washington, D.C. on March 1, 1975, by the "March First Coalition."[50]

Meanwhile back home in the US, people protested against the war in the streets, on college campuses, and in Washington, DC. Some who were drafted easily traveled to other countries like Canada to avoid serving in the military. I remember in college men telling me:

"I am getting a teacher's job to avoid the draft."

"I am getting a defense job."

"Better red than dead."

"My doctor will give me a note that says I have xyz ailment." Even though I am physically very active.

One former President joined the National Guard and hardly served there. Another got a note from his doctor stating that he had bone spurs.

---

[49] Kent_State_massacre.jpg (300 × 238 pixels, file size: 19 KB, MIME type: image/jpeg) https://en.wikipedia.org/wiki/File:Kent_State_massacre.jpg . On publication, the image was retouched to remove the fencepost above Vecchio's head.
[50] Seattle Municipal Archives from Seattle, WA - Seattle Municipal Archives, Public Domain, https://commons.wikimedia.org/w/index.php?curid=35019367

None of these guys had any political sentiments about the war. They were just against someone else having authority over them or they didn't care about other races becoming communist, or they just didn't want to go.

However, others did have political and moral feelings about the war. At the start of the protests a... "vocal minority included many students as well as prominent artists, intellectuals and members of the "hippie" movement, which was the growing number of mostly young people who rejected authority and embraced the counterculture."[51]

"By November 1967, American troop strength in Vietnam was approaching 500,000 and U.S. casualties had reached 15,058 killed and 109,527 wounded. The Vietnam War was costing the United States some $25 billion per year, and disillusionment was beginning to reach greater sections of the taxpaying public.... 40,000 young men were called into service every month, adding fuel to the fire of the antiwar movement."[52]

There have been songs of protest throughout history and the unpopularity of the Vietnam War brought forth a whole slew of new ones. In the early days of the 1960s, many popular songs such as "Where Have All the Flowers Gone?" by the Kingston Trio were objections to the futility of war in the wake of the scare of the Cuban Missile Crisis. But as the Vietnam War picked up steam and began reaching into the homes of middle America, the songs became more specific in targeting not just war in general, but the Vietnam War in particular.

In the mid 1960s Phil Och's "I Ain't Marching Anymore" and Richie Havens' "Handsome Johnny" were protests against Vietnam. As the decade progressed, the songs became more strident and politicized as the singer/songwriters poured their anger at the whole concept of the War into their music like "Long Time Gone" by Crosby, Stills and Nash (inspired by the assassinations of MLK and RFK) and "Ohio" inspired by the killings at Kent State. Even the Beatles wanted "Revolution" and the Rolling Stones sang of "Street Fighting Man," songs all born from the tumult and turbulence of the times.

By 1969 John Lennon was pleading, "Give Peace a Chance." Later he wrote "Imagine," a song of peace and unity which became a classic. There were

---

[51] https://www.history.com/topics/vietnam-war/vietnam-war-protests
[52] https://www.history.com/topics/vietnam-war/vietnam-war-protests

many other "…popular songs that became an anthem for that generation. Phil Ochs wrote "What Are You Fighting For?" in 1963 and "I Ain't Marching Anymore" in 1965…Pete Seeger's "Bring 'Em Home" (1966), Creedence Clearwater Revival's, "Fortunate Son", says the poor didn't get out of the war, while the well-connected sons were safe at home.[53]

The organization, "*Vietnam Veterans Against the War*, many of whom were in wheelchairs and on crutches, joined the protests. The sight of these men on television throwing away the medals they had won during the war did much to win people over to the antiwar cause."[54]

Protests While I was in Vietnam 70-71

As the attitude of the US and the Europeans changed against the war, large scale protests became more prevalent. On April 24, 1971, in Washington, DC there were anti-war protests. People carried posters saying "*Veterans* and *WAR SHITS – OUT NOW*"[55]

Figure 1-13 Anti-war protest against the Vietnam War in Washington.

One of the largest was the May Day 1971 protest in Washington, DC. Which occurred at the same time I commanded an engineer company in Vietnam. These protests created doubt in many of the soldiers. However, they all did their jobs and soldiered on.

Those protests were a series of large-scale civil disobedience actions in Washington, D.C. They began on Monday morning, May 3rd, and ended on May 5th. More than

---

[53] https://www.history.com/topics/vietnam-war/vietnam-war-protests
[54] https://www.history.com/topics/vietnam-war/vietnam-war-protests
[55] Photo by Leena A. Krohn. This file is licensed under the Creative Commons Attribution-Share Alike 3.0 Unported license. https://commons.wikimedia.org/wiki/File:Vietnam_War_protest_in_Washington_DC_April_1971.jpg

12,000 people were arrested, the largest mass arrest in U.S. history. The Federal Government called out the US Park Police, The DC Police, the 82nd Airborne Division, The U.S. National Guard, and the U.S. Marine Corps. Approximately 19,000 Federal troops, National Guardsmen, and local police were involved.

More than 40,000 protesters camped out in West Potomac Park near the Potomac River to listen to rock music and to plan for the coming action.

Members of the Nixon administration would come to view the events as damaging, because the government's response was perceived as violating citizens' civil rights. [56]

There were also large scale protests in Europe.

So, the soldiers continued to ask, "Why are we here? Does anybody care?"

Jane Fonda was an extremely famous movie actress, environmentalist, and political activist. She vigorously led protests against the Vietnam War. Because of her status, the anti-Vietnam War efforts gained significant momentum.

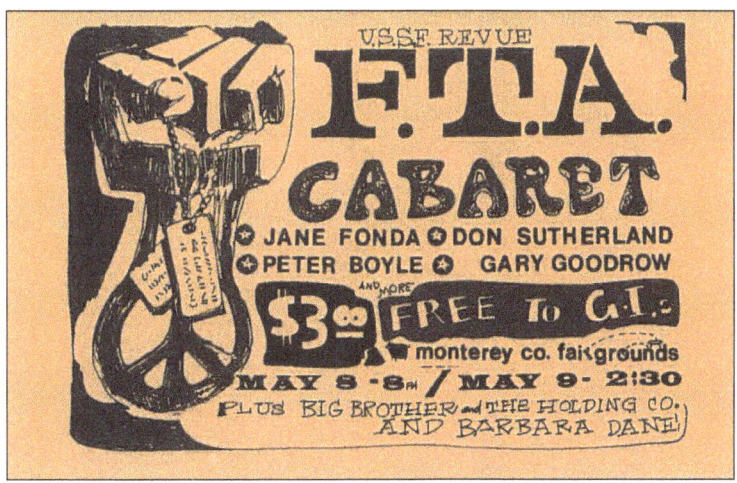

Figure 1-14 Flyier for Jane Fonda's Fuck the Army (FTA) rally

She and other antiwar activists such Paul Newman, Dr Spock, the Smothers Brothers, Walter Cronkite, and others actively spoke out against the war. She

---
[56] https://en.wikipedia.org/wiki/1971_May_Day_protests

helped organize a troupe that would provide entertainment to U.S. soldiers who opposed the war in Vietnam. The troupe, called FTA (Fuck The Army), performed songs, comedy sketches, and dramatic readings. The show toured outside military bases in the United States and for GIs stationed abroad in the Pacific Rim. Other celebrities and entertainers joined Fonda for FTA's including actor and activist Donald Sutherland (best known for his role in the satirical film *MASH*, 1970), comedian and civil rights activist Dick Gregory, and folksinger and longtime activist Barbara Dane.[57]

Unknown to most people, between 1965 and 1972, almost 300 Americans – mostly civil rights activists, teachers, and pastors – traveled to North Vietnam to see firsthand the war situation with the Vietnamese. News media in the United States predominantly provided a U.S. viewpoint, and American travelers to Vietnam were routinely harassed upon their return to the States. Jane Fonda also visited Vietnam.[58]

She made radio broadcasts on Hanoi Radio throughout her two-week tour, describing her visits to villages, hospitals, schools, and factories that had been bombed, and denouncing U.S. military policy. Notice in the picture the sign on the right. It says, "JANE FONDA Indo-China Peace Campaign." During the course of her visit, Fonda visited American prisoners of war (POWs), and brought back messages from them to their families. When stories of torture of returning POWs were later being publicized by the Nixon administration, Fonda said that those making such claims were "hypocrites and liars and pawns," adding about the prisoners she

Figure 1-15 Jane Fond at Press Conference in Holland

---

[57] https://ushistoryscene.com/article/fta/ This image can be found at http://peoplesoralhistoryprojectmc.org/historical-photos-of-activism-monterey-county/ courtesy of Corey Miller. This ticket is from FTA's performance in Monterey, California (from Peoples Oral History Project Monterey County).

[58] https://upload.wikimedia.org/wikipedia/commons/thumb/3/3b/Jane_Fonda_1975d.jpg/1024px-Jane_Fonda_1975d.jpg

visited, "These were not men who had been tortured. These were not men who had been starved. These were not men who had been brainwashed." In addition, Fonda told The New York Times in 1973, **"I'm quite sure that there were incidents of torture ... but the pilots who were saying it was the policy of the Vietnamese and that it was systematic, I believe that's a lie."** (My emphasis). Her visits to the POW camp led to persistent and exaggerated rumors which were repeated widely and continued to circulate on the internet. Decades later, Fonda, as well as named POWs, have denied the rumors, and subsequent interviews with the POWs, showed these allegations to be false—that the persons named had never met Fonda. [59]

In a 2005 interview on 60 Minutes, she said that, "…she had no regrets about the broadcasts she made on Radio Hanoi, something she asked the North Vietnamese to do. "Our government was lying to us and men were dying because of it, and I felt I had to do anything that I could to expose the lies and help end the war."[60]

The picture shows Jane Fonda seated on a North Vietnamese anti-aircraft gun, which earned her the nickname "Hanoi Jane" because she sided with

Figure 1-16 Jane Fonda seated on a North Vietnamese anti-aircraft gun; and earned her the nickname "Hanoi Jane"

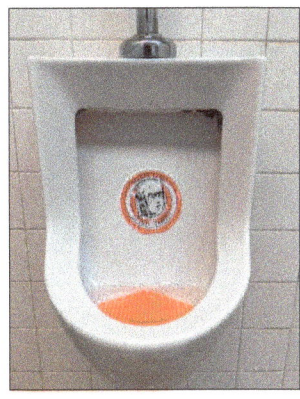

Figure 1-17 Jane Fonda urinal target in some American Legion Post restrooms.

---

[59] Nederlands: Jane Fonda geeft persconferentie in TH in Delft voor Vietnamacties; Jane Fonda tijdens persconferentie. 17 January 1975 [1] Dutch National Archives, The Hague, Fotocollectie Algemeen Nederlands Persbureau (ANeFo), 1945-1989, Nummer toegang 2.24.01.05 Bestanddeelnummer 927-6990 Mieremet, Rob / Anefo This file is licensed under the Creative Commons Attribution-Share Alike 3.0 Netherlands license. https://commons.wikimedia.org/wiki/File:Jane_Fonda_1975d.jpg
https://en.wikipedia.org/wiki/Jane_Fonda#Visit_to_Hanoi
[60] https://en.wikipedia.org/wiki/Jane_Fonda#1970_arrest

the North Vietnamese. Many soldiers had hard feelings against her considering her a traitor. So much so, that even today urinals in some American Legion posts have an image of her. After all, she was pictured sitting on an anti-aircraft gun. However, in her 2005 autobiography, she wrote that she was manipulated into sitting on the battery; she had been horrified at the implications of the pictures.[61] She said, "It is possible that it was a set up, that the Vietnamese had it all planned. I will never know. But if they did I can't blame them. The buck stops here. If I was used, I allowed it to happen... a two-minute lapse of sanity that will haunt me forever... But the photo exists,[62] delivering its message regardless of what I was doing or feeling. I carry this heavy in my heart. I have apologized numerous times for any pain I may have caused servicemen and their families because of this photograph. It was never my intention to cause harm."[63]

Nevertheless, her actions encouraged Ho Chi Minh, the North Vietnamese Army, and the Viet Cong to believe that they were winning the war and to continue their struggle. Consequently, she indirectly killed and maimed many US and South Vietnamese soldiers.

Jane Fonda's actions were a blow to many soldiers. She helped foster the feeling, "Does anybody care?"

* * * * *

Bui Tin, a former colonel in the North Vietnamese army, answered questions in an interview conducted by Stephen Young, a Minnesota attorney and human-rights activist. Bui Tin, who served on the general staff of North Vietnam's army, received the unconditional surrender of South Vietnam on April 30, 1975. He later became editor of the official newspaper of Vietnam. Later he moved to Paris, where after becoming disillusioned with Vietnamese communism.

He said that, "Hanoi intended to defeat the Americans by fighting a long war which would break the American will to fight and to help South Vietnam. Furthermore, he said, that Ho Chi Minh said that American antiwar

---

[61] https://en.wikipedia.org/wiki/File:Hanoi_Jane.jpg
[62] https://www.govinfo.gov/content/pkg/CREC-1995-08-04/html/CREC-1995-08-04-pt1-PgH8514.htm
[63] https://en.wikipedia.org/wiki/Jane_Fonda#Visit_to_Hanoi

movement was important to Hanoi's victory. That it was essential to our strategy. Support for the war from our rear was completely secure while the American rear was vulnerable. Every day our leadership would listen to world news over the radio at 9 a.m. to follow the growth of the American antiwar movement. Visits to Hanoi by people like Jane Fonda and former Attorney General Ramsey Clark and ministers gave us confidence that we should hold on in the face of battlefield reverses. We were elated when Jane Fonda wearing a red Vietnamese dress, said at a press conference that she was ashamed of American actions in the war and that she would struggle along with us".

Or in my words, "Just kill and maim enough GI's to appear on the nightly TV news that the American public will tire of the war." The protestors thought they were saving lives. Many GI's thought that the protestors were aiding the enemy and killing GI's.

Consequently, the soldiers in my unit were aware of these protests which bothered them. They continually asked me, "Why are we here?"

I usually responded, "Let's do our jobs and go home."

Inside I felt, "This is a lousy way to conduct a war. Who wants to be injured, or maimed, or killed for a war the country, the political and religious leadership, and the public does not want? Why aren't the President and Congressional leaders talking to us soldiers?"

Race Riots and Marches

This section has been the most difficult of the whole book for me to write. In part because I was very unfamiliar with it. Also, in part because without realizing it, prior to my experience in Vietnam from 1970-1971, I had very little contact with racial issues. Also, as a white company commander in charge of a unit consisting of a majority of white soldiers I found that the African American soldiers and I kept our distance from one another. Whether this was because of my ignorance or built-in bias I don't know.

Now, fifty-three years later I am finally thinking about one of the major problems in the unit which was race relations, some of the actual events I describe later in this book. In this section, I am attempting to understand the historical context of race relations in the US which then affected the unit.

Several people mentioned to me that I should not include the racial riots and Martin Luther King's comments about the Vietnam War because I did not know if this influenced the African American soldiers in my unit. Some people emphatically mentioned, "You don't know if any of those events affected them in Vietnam. It is poor scholarship to include this material if you don't know it affected them or even if was significant to them." This is technically a fair statement since I did not discuss these issues with any of them neither during nor after the war. Yet, I can't help but feel that the attitudes of the African American soldiers were influenced by their treatment by the White majority both before and during the war. As for myself, I was vaguely aware of race riots, racism, slavery, and the history of minorities in the United States. However, I had previously not taken much notice of it until the war when I was forced to confront it head on.

Even though my wonderful eighth grade teacher, Ida Powell in 1960, mentioned the ideals of the US. Declaration, "We hold these truths to be self-evident, that **all men are created equal, that they are endowed by their Creator with certain unalienable Rights, that among these are Life, Liberty and the pursuit of Happiness**.--That to secure these rights, Governments are instituted among Men...."She then skillfully led discussions about what that meant in the US. She also mentioned that these are our national ideals and that sadly we haven't and don't live up to them as a country. However, we can decide to make changes to live up to those ideals.

She then mentioned the Tulsa riots, the Wilmington riots, and conditions of slavery ranging from benevolent to vicious. Furthermore, she delicately described lynching of African Americans as more than several people grabbing an African American man and hanging him in some remote forest. In a number of cases it was an form of entertainment and at times was brutal. Though she didn't elaborate on it.

As a fourteen year old eighth grader, I was incredulous. Would people really treat others like that? That is impossible.

The closest I'd come personally to any racial issue took place about two years later during a summer recess from high school when I went to stay with family friends in Virginia for several weeks.

The husband worked at a bank. So, he and his wife took me along to the bank party being held on a boat. It was a delightfully fun time. The boat set sail down a river. The jokes, the laughter, and free flowing conversation were a true joy. Then in the darkness, one of the men yelled out, "Here comes the African Queen!"

Suddenly, as if they were competing in a synchronized swimming event, all the men swarmed to the side of the ship. They began yelling and screaming into the deep velvet blackness. But, I couldn't see anything. What were they yelling at? Vaguely, I began to make out a large multistory vessel that was beginning to pass the one deck party boat. Dimly I perceived that the ship had four decks with African American people seated on benches facing outwards enjoying the sights.

The men on the boat I was on grabbed empty beer and wine bottles and threw them at the ship while yelling, with their faces flushed red. While the party boat was well lit in the darkness, the other ship had only running lights in the front, at the top, and behind the ship. Those people almost vanished in the blackness.

I was scared that all those hundreds of people would soon be throwing bottles back at us. I searched for a place to hide. The women said, "Oh come on guys. We want to have fun. They want to have fun. Just leave them alone."

As the ship passed, one woman's strong voice from the ship pierced the darkness and the noise with, "We shall overcome." Then a few more began to sing, "We shall overcome." Then more. Then the whole ship began to sing, "We shall overcome. We shall overcome some day. Deep in my heart, I do believe, we shall overcome some day." Over and over they sang as the men on the party boat continue to yell and scream.

The ship powerfully and steadily continued on past the party boat fading into the darkness. The strong voices all in unison continued to sing. Their vigorous voices became fainter yet pierced my soul, deep into my soul. How could they be so calm and so strong in the face of such hatred?

Little did I know that years later in the Vietnam War that I would be in the middle of the same type of scene. The difference being that I was unknowingly in danger of being seriously hurt or even killed. That story comes later.

* * * * *

From 1910 to 1970, African Americans fled the South to escape the terror of lynchings, a historic event known as the Great Migration.[64] People began to oppose lynchings in a number of ways. They conducted grassroots activism, such as boycotting white businesses. Anti-lynching crusaders like Ida B. Wells composed newspaper columns to criticize the atrocities of lynching.

And several important civil rights organizations — including NAACP — emerged during this time to combat racial violence.

Figure 1-18 A 17 year old Jesse Washington lynched. Note the crowd of smiling spectators posing for the camera. July 1916.

Figure 1-19 Crowd proudly forms for picture of lynching. Note the legs above the crowd. August 3, 1920.

The NAACP led a courageous battle against lynching. In the July 1916 issue of The Crisis, editor W.E.B. Du Bois published a photo essay called "The Waco Horror" that featured brutal images of the lynching of Jesse Washington.[65]

Washington was a 17-year-oteen lynched in Waco, Texas, by a white mob that accused him of killing Lucy Fryer, a white woman. Du Bois was able to turn postcards of Washington's murder against their creators to energize the anti-lynching movement. The Crisis's circulation grew by 50,000 over the next two years, and we raised $20,000 toward an anti-lynching campaign.[66]

In 1919, NAACP published Thirty Years of Lynching in the United States, 1889-1919, to promote

---

[64] https://www.archives.gov/research/african-americans/migrations/great-migration
[65] https://naacp.org/sites/default/files/styles/embed_image_c/public/images/lynchingofjessewashington.webp?itok=V9vZKfBb
[66] https://naacp.org/find-resources/history-explained/history-lynching-america

awareness of the scope of lynching. The data in this study offer the gruesome facts by number, year, state, color, sex, and alleged offense.

Men and boys pose beneath the body of an African American man, shortly after he was lynched on August 3, 1920, in Center, Texas.[67]

Figure 1-20 Sign at bus station.

However, there were daily reminders such as they must site in their own areas, nor sit in the front of buses, and give up their seats when a white person came on the bus.[68]

There were restaurants, stores, drinking fountains, restrooms, grocery stores, etc. that posted signs, "Whites only." When traveling the Black person used what was called the Green Book. These were hotels and motels that accepted Black people so that they could plan their overnight stops. Otherwise, they would spend the night in their cars. It was a common experience for the police to stop black people while driving or walking just to harass them and never charge them with a crime. In some cases, a white woman or man would accuse a black man of 'looking leeringly at a white woman.' That would be considered a crime.

However, there were individuals and communities that thrived and became economically successful and had their own politicians. Yet, in different ways, hatred pulled some of those communities apart. For example, Whites destroyed the black communities of Wilmington, North Carolina and a section of Tulsa, Oklahoma. Furthermore, in other communities such as Asheville, North Carolina, a highway was built right through the community. Unfortunately, the forced integration of blacks and whites created white flight and destroyed some very successful black schools. With integration black students were mixed with white students. As the schools consolidated, the male and female black teachers who were role models for their students were frequently not hired. The white teachers either purposefully or inadvertently looked down upon the black students.[69]

While well meaning, government programs that were meant to help the black community such as welfare laws, encouraged the breakup of the

---

[67] https://lynchinginamerica.eji.org/report/
[68] http://loc.gov/pictures/resource/cph.3b22541/ No known restrictions.
[69] https://www.exploreasheville.com/listings/hood-huggers-international-tours/8132/

black family. Nevertheless, there were many successful role models such as Oprah Winfrey.

Personally, however, I was barely aware of such daily occurrences. One time my mother and I when I was fourteen years old were visiting a good friend of hers in Virginia. We went out for a nice dinner. No one came to ask us for drinks nor give us a menu. My mother motioned to a waitress to come over and quickly said to me, "Don't say anything!"

Then a waitress came over and said, "We don't serve that person here. She has to leave."

My mother said to her, "She is from Costa Rica."

The waitress said, "Oh, that is okay. We just want to make sure that those people don't get ideas and start coming in here."

We were served.

Figure 1-21 Book Cover "Manchild in the Promised Land"

That same year in my high school English class, the teacher required us to read, "Manchild in the Promised Land" by Claude Brown.[70] Published in 1965 during the Civil Rights Movement, it described a child in Harlem struggling to survive violence and poverty in Harlem, New York City. Yet, I was so focused upon surviving the engineering and math classes, that story felt like science fiction. Nevertheless, it made a lifelong impression upon me.

Such was my understanding of the Black situation when I would later assume company command and be constantly puzzled by the low level constant racial friction that would later erupt into a near race riot and another instance that resulted in an assault upon one of my lieutenants. Even today, while I am much more aware of the racial issues, the problem still puzzles me. Why do people hate each other?

---

[70] https://en.wikipedia.org/wiki/Manchild_in_the_Promised_Land#/media/File:ManchildInThePromisedLand.jpg "qualifies as fair use under the copyright law of the United States."

On July 26, 1948, President Harry Truman signed Executive Order 9981, creating the President's Committee on Equality of Treatment and Opportunity in the Armed Services. The order mandated the desegregation of the U.S. military. The first point in the executive order states "It is hereby declared to be the policy of the President that there shall be equality of treatment and opportunity for all persons in the armed services without regard to race, color, religion or national origin. This policy shall be put into effect as rapidly as possible, having due regard to the time required to effectuate any necessary changes without impairing efficiency or morale."[71]

Figure 1-22 Members of the 2nd Inf. Div. north of the Chongchon River. Sfc. Major Cleveland, weapons squad leader, points out a North Korean position to his machine gun crew in 1950.

Truman's order received pushback from politicians, generals, and friends, who opposed an integrated military. Truman wrote in response to his detractors, "I am asking for equality of opportunity for all human beings, and as long as I stay here, I am going to continue that fight."

The African American soldiers in my unit came of age during the marches, riots, and beatings in the 1960's and the early 1970's. They also probably heard stories of the Tulsa, Oklahoma, the Wilmington riots in the 1920's and 1930's, and many other places, where African American businesses, homes, and communities were destroyed by whites, as well as Truman's desegregation of the US military.

Therefore, my guess is that African American soldiers were keenly aware of as many as 700 civil disturbances between 1964 and 1971 which created large numbers of injuries, deaths, and arrests, as well as the significant property damage which was concentrated in predominantly African American areas.

African American communities struggled to gain the normal political, social, educational, and financial rights enjoyed by many white Americans. Marches and riots were attempts to change the fundamental structure of American society for equality for African Americans and all races. As the

---

[71] https://www.nps.gov/articles/000/executive-order-9981.htm

Reverend Martin Luther King stated, in August 1963 march in Washington, D.C. "I have a dream that my four children will one day live in a nation where they will not be judged by the color of their skin, but by the content of their character."[72] In 1965 he stated, "How long will this take? How long?"[73]

Figure 1-23 March on Washington 1963 where the "I have a Dream Speech" was given.

Meanwhile, the Reverend Martin Luther King advocated for non-violence using peaceful marches and political actions.[74] Eventually, President Johnson signed the Civil Rights Act of 1964.[75] Martin Luther King stands behind President Johnson. Johnson later stated the Democrats would lose the South.

Figure 1-24 President Lyndon B Johnson signs the 1964 Civil Rights Act as Martin Luther King, Jr., and others, look on. July 2, 1964.

Another example that had a major influence on the soldiers was the Detroit Riot of 1967. It was one of the largest and began on July 23, 1967, and lasted five days. 43 people died, including 33 African Americans and 10 whites. Many other people were injured, more than 7,000 people were arrested, and more than 1,000 buildings were burned in the uprising.

The riot is considered one of the catalysts of the Black-Power movement.[76]

---

[72] https://www.npr.org/2010/01/18/122701268/i-have-a-dream-speech-in-its-entirety
[73] https://voicesofdemocracy.umd.edu/dr-martin-luther-king-jr-long-not-long-speech-text/
https://news.harvard.edu/gazette/story/2013/08/the-dream-50-years-later/
[74] https://upload.wikimedia.org/wikipedia/commons/7/7f/March_on_washington_Aug_28_1963.jpg
[75] https://www.nps.gov/gwmp/learn/historyculture/images/LBJ-signs-Civil-Rights-Bill-MLK-and-others-stand-behind-him.png
[76] Emeka, Traqina Quarks. "Detroit Riot of 1967". *Encyclopedia Britannica*, Invalid Date, https://www.britannica.com/event/Detroit-Riot-of-1967. Accessed 3 May 2021.

Hopefully, this short section gives a sense of the emotions that African Americans and some whites would have experienced and/or would have been told by their parents, relatives, friends, and others by the time they appeared in my company.

Figure 1-25 Detroit Riot 1967 and 7,000 arrested

I, however, was barely aware of any of this. Images like these and many more appeared in the hardcopy press and on national television. Consequently, American opinion and feelings on race changed like a creeping glacier. Nevertheless, while improvements were being made, the underlying prejudices remained in more subtle ways and continue on to today.

I think what I just described was deeply embedded to the very blood and bones of the African American and White soldiers in my unit. And there were probably many who were as ignorant and naive about these issues as I was.

However, I did not take sides in their fight for Civil Rights; instead, I focused on accomplishing the mission, and getting everyone home unharmed. Perhaps I should have attempted to educate everyone about racial equality with as much effort as I did about safe sex, avoiding drugs, and getting the job done. But I was instinctively afraid of creating problems with race relations that would adversely affect getting the job done. The way around it was to keep everyone working so that all they cared about was sleep, food, getting laid, and going home. That way, the racial problems would be minimized. I did not like avoiding the racial problems. I just had to figure out how to deal with the simmering conflict and accomplishing the mission.

Nevertheless, in writing this section, I took into consideration that some people strongly encouraged me to not include the graphic detail. However, I feel that to have omitted it would hide the background of the African American experience that soldiers in my unit were probably aware of. That to lightly touch on the race issue would be burying a major problem that had occurred in several units in the US Army including the ones I confronted and was occurring in the US during the Vietnam War. So, I have chosen to include snippets of a long and troubled history so that the reader can get a sense of the context. More fundamentally for the reader to understand another issue I regrettably had little knowledge or familiarity with. Yet, as will be related later, I had to deal with many times and in one instance could have been injured or killed. Sadly, these underlying subtle prejudices have continued for another fifty years.[77]

So, I repeatedly quietly said, "Let's just do our jobs and go home."

Drug Usage

"The image of the drug-addicted American soldier—disheveled, glassy-eyed, his uniform adorned with slogans of antiwar dissent—has long been associated with the Vietnam War and has been given as a reason for the US defeat. It became a symbol of a demoralized army incapable of carrying out its military mission. However, Jeremy Kuzmarov in his book, *The Myth of the Addicted Army*, builds a case that drug usage in Vietnam is based more on myth than fact. Not only was alcohol the intoxicant of choice for most GIs, but the prevalence of other drugs varied enormously. Marijuana use among troops increased over the course of the war, was confined to rear areas, and the use of highly addictive drugs like heroin was never as widespread as many imagined."[78]

Figure 1-26 The Myth of the Addicted Army made the case that drug addiction was not widespread in the US Army.

The concept of an addicted army was first advanced by war hawks seeking a scapegoat for the failure of US policies in Vietnam. Some blamed liberal social policies and the excesses of the counterculture. Also, Kuzmarov states, elements of the antiwar movement promoted the myth. They presumed an alliance between Asian drug traffickers and the Central Intelligence Agency. While this claim was not without foundation, the left exaggerated the scope of addiction for its own political purposes demonstrating that the US Army was incapable of fighting a war. Therefore, another reason to go home.

---

[77] https://www.nps.gov/gwmp/learn/historyculture/lbjandmlk.htm
[78] https://www.amazon.com/Myth-Addicted-Army-Vietnam-Politics/dp/1558497056

Exploiting bipartisan concern over the perceived "drug crisis," the Nixon administration in the early 1970s launched the "War on Drugs" which enjoyed broad bipartisan appeal. Public officials from both the left and the right were quick to blame marijuana and heroin for American failures abroad. Democratic Senator Thomas J. Dodd claimed illegal drug use directly contributed to the Mai Lai massacre and other American atrocities of war, stating that, "Tens of thousands of troops have gone into battle high on marijuana, opium or other drugs, with horrifying results."[79] Thus it helped divert attention away from the failed quest for "peace with honor" in Southeast Asia. But once institutionalized, it continued to influence political discourse as well as U.S. drug policy in the decades that followed"[80]

There was some public concern that drug addicted soldiers would return from Vietnam and abuse drugs at home. In response to that anxiety, the White House implemented "Operation Golden Flow" in 1971, which mandated that all servicemen subject themselves to urinalysis before boarding planes back to the United States. Should a serviceman fail to pass his drug test, he was required to stay in the country for detoxification, only to be released back to the United States upon successfully testing clean.

Anxiety about mass addiction returning to America's shores proved misplaced. Whether a result of Operation Golden Flow or a sign of the more casual usage than initially reported, an interview survey commissioned by the White House's Special Action Office for Drug Abuse Prevention found that usage and addiction rates "essentially decreased to pre-war levels" following the soldiers' return.[81]

Nevertheless, 48% of Americans in a Gallup poll, in 1969, stated that drug use was a serious problem in their community.[82] It was a significant problem with soldiers in my unit. Sadly, I was not able to solve the problem, help them, and still get the mission accomplished.

Today, drug addiction continues to be a serious problem. The US government, politicians, the medical community, businesses, education, and religious organizations, have failed to minimize the problem.

---

[79] https://www.history.com/news/drug-use-in-vietnam
[80] https://www.amazon.com/Myth-Addicted-Army-Vietnam-Politics/dp/1558497056
[81] https://www.history.com/news/drug-use-in-vietnam
[82] https://news.gallup.com/poll/6331/decades-drug-use-data-from-60s-70s.aspx

With no vision, the people perish

When a war spans several Presidential administrations there is a serious risk of losing the vision and the will of the American people. This is a fundamental weakness of the American political system. Especially when the enemy is a dedicated, entrenched political force, led by one dictator who knows enough to not arouse the American spirit. Rather the dictator will nibble, nibble, and nibble while the American Presidents and leadership dream about staying in office and not upsetting the American public.

The various dictators in the world view their countries as strong and unified; while they consider democracies consist of corrupt capitalism and lack traditional morals, both of which enfeebles the West. That eventually their strong system will win and the incoherent West's system will collapse. Their view emboldens them to take aggressive actions against a feminine and decadent West. Actions such as President of Russia, Vladimir Putin's invading Ukraine and President of China, Xi Jinping, planning to invade Taiwan. And so many other autocrats in the world.

This is the problem the US and Western Europe face with Russia and China. Both Russia and China have grand visions of great empires and missions for the world, while the US and Europe as societies lack an effective vision for the world. I am confident that various political leaders have their visions, but the US and Europe lack a coherent view when contrasted with Russia and China. Proverbs 29 verse18 states what I just described. "Where there is no vision, the people perish."[83]

The lack of leadership by Presidents Kennedy, Johnson, Nixon, and Ford, as well as their administrations to sell the Vietnam War to the American public and the world, resulted in antiwar protests, race riots and marches, lies by the various administrations, as well as increasing drug usage at home and in Vietnam. All of which encouraged the Viet Cong and the North Vietnamese.

Consequently, I quietly said over and over to soldiers, "Let's just do our jobs and go home."

---

[83] Though I purposely omitted the last part of verse 18, "but he that keepeth the law, happy is he." Adding the phrase would change the nature of the discussion which I would gladly get into. Nevertheless, I still think the first part of the verse has an appropriate place here.

The Result

The result was that the soldiers were not in Vietnam for the duration of the war. Unlike WWII there was literally no effective moral nor emotional support for them. However, the Army did an excellent job providing physical comforts for most of the soldiers. Though I am sure there were many in the so-called front lines pounding around the jungle that wouldn't agree. There were major installations in Saigon (renamed Ho Chi Minh City after the US pulled out), Da Nang and other places that gave the feeling of being in the US on the military base. Though there was a sense an attack could happen at any time. However, the small company size compounds were subject to attacks or just harassment.

Consequently, because of the lack of emotional, political, religious, and family support, many soldiers created their own countdown calendars until they would go home. They had expressions like, 205 days and a wakeup.

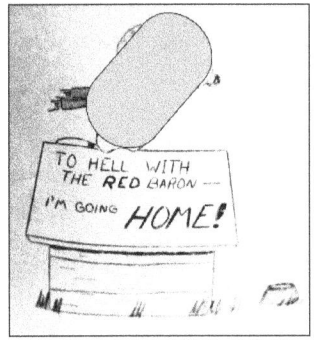

Meaning that they had 205 days left and then they would be on the plane that they named the "Freedom Bird."

This was a lousy way to fight a war. On the one hand the military was highly concerned about winning the minds and hearts of the Vietnamese; yet on the other hand, there appeared to be little concern about winning the minds and hearts of the average soldier. It was one of many fundamental flaws of the war.

Figure 1-27 Many soldiers carried Snoopy with their count down calendars. It showed their main hope – go home.

Because of copyrights Snoopy is blurred out.

It is a similar problem now. The US public is divided. It is physically and emotionally soft; too soft to carry on a brutal war that will span decades. While we rely on technology to fight, we as a society don't have the will. Though there are many who parade around the streets with various pistols and rifles feeling they are supporting the US, they are mostly fat, unfit, and unable to sustain protracted war. As Lieutenant General Mark Hertling said, "It is the US's Achilles heel.[84]

---

[84] https://mwi.usma.edu/mwi-podcast-physical-fitness-national-security-lt-gen-ret-mark-hertling/

As the source noted in the footnote states, "The military depends on a constant flow of volunteers every year. According to 2017 Pentagon data, 71 percent of young Americans between 17 and 24 are ineligible (because of being physically unfit, drug usage, criminal record, mentally not qualified) to serve in the United States military…. Over 24 million of the 34 million people of that age group cannot join the armed forces—even if they wanted to. This … threatens the country's fundamental national security. If only 29 percent of the nation's young adults are qualified to serve, and if this trend continues, it is inevitable that the U.S. military will suffer from a lack of manpower. A manpower shortage in the United States Armed Forces directly compromises national security."[85]

Unlike the Japanese attack on Pearl Harbor which galvanized the country, no US President nor politician worked to galvanize US and international support for the Vietnam War. Instead there was a lot of behind the scenes maneuvering to either get out of the war, win the war, or to negotiate a peace with honor. However, because the North Vietnamese leadership publicly advocated for their War of Independence from at least 1945 to 1975.

Consequently, with such long term political work, large segments of both the North and South Vietnamese populations supported their war for independence. They were conducting war as described by Mao Zedong in his book, *On Protracted War*.[86] It was and is *the* key book about successfully conducting guerilla warfare and if necessary to conduct it for decades if not even longer.

Thus, creating the political will for protracted war for decades is an effective way to win. This is especially true when the oppositioin lacks such stern discipline and will to fight a protracted war. As Will and Ariel Durant wrote in their eleven volume set, *The Story of Civilization*, 'Wars frequently end when one or both sides are exhausted and they just go home.'

---

[85] https://www.heritage.org/defense/report/the-looMinh-national-security-crisis-young-americans-unable-serve-the-military
[86] Mao ZeDong, *On Protracted War,* 1938. Written for combatting the Japanese from 1938 to 1945. Rapidly used to combat the French and the Americans from 1945-1975. Thus, it was a guiding principle for nearly forty years of war.

"Good Idea Son"

A major problem with both the US and its allies is their inability to politically, physically, emotionally, economically, and logistically conduct a protracted war against China, Russia, North Korea, and Iran at the same time. The West has the technology, but does it have the intestinal fortitude to fight a protracted war for fifty or more years so that there is economic and social freedom in both the US and the world? Sadly, totalitarian governments seem to convince their people to sacrifice to conduct protrated wars.

Thus, China is forming a dictatorial world order "with Chinese characteristics". The competing world is fragmented with one part tolerating everything and the other leaning towards theocratic fascism. China is prepared to play the long game and to bide its time for more power like a hunter using traps. Thus, fascist, theocratic, and Chinese leaders are quietly stalking their prey.

Twenty years after my part in the War, I was at a fund raiser for President George H.W. Bush. After dinner he mingled with the crowd. As he passed me, I called out, "Mr. President!"

He turned towards me. Suddenly, I felt a big goon behind me on the left and another one behind me on the right. Then the President was standing six or so feet in front of me. There were two big guys to his left and two to his right.

As I glanced around several times at the big guys, I blurted out, "I forgot what I was going to say."

The Secret Service guys and Bush laughed. As they turned to move on I said, "I remember what I was going to say." They stopped and again automatically formed around me.

I said to him, "The next time you have a war, get the support of the American people and the world. Let the Generals and the Admirals fight the war."

He made a thoughtful gesture with his left hand to his chin and nodding said, "Good idea son!" Turned and continued walking through the crowd. I knew then that he already knew what I had just said. He was just being polite.

Since then, I have jokingly said that it was my idea that he lined up the US Congressmen and Senators to publicly sign a document supporting what became known as the First Gulf War. That way he coherently gathered support from the US population and many in the world to support the War effort. Sadly, no President gathered that kind of support for the Vietnam War, which was a fundamental gross political failure.

One might quibble that President Kennedy did try to gain public support; however, he was assassinated so those initial efforts vanished at the barrel of gun. Perhaps the whole war effort would have changed had he not been assassinated. Or, it could have failed anyway because of other reasons explained earlier in this chapter. Or, most likely because he was realizing the futility of conducting such a war, maybe he would have pulled out before committing the US to it.

The point being, it's vitally important for any country and its allies to gain the trust and support of the population before committing them to going to War. When I worked in China on and off from 2005-2008 it was evident that the Chinese leadership was campaigning to win the hearts and minds of the Chinese people to take over Taiwan. They were advertising a growing navy with an aircraft carrier, etc. The US leadership is a blurred mixture of the Left versus the Right. The result is a lack of unified vision in the US.

Summary

This was the context of the Vietnam War when I assumed company command. Events in the US and around the world affected many of the soldiers in the unit and I am sure many of the soldiers in Vietnam. We were just average people caught up in a war that was ineffectively led at the highest levels of the US and South Vietnamese governments.

The US and Vietnamese political leaders stole lives from the Vietnamese and Americans who died or were maimed there, from relatives quietly mourning in villages and homes across the US and Vietnam, as well as a year of life from every American, South Vietnamese, and North Vietnamese soldier who served there.

So, without fully realizing the problems and issues described in this chapter, rightly or wrongly, I said to my soldiers, "Let's do our jobs and go home."

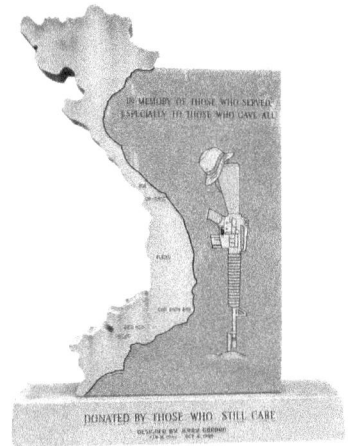

So, we did our jobs,

we went home,

and nobody cared

Chapter 2
Welcome to Vietnam

Orientation Training at Fort Lewis, Washington

Prior to arriving in Vietnam, the Army required one week of orientation training. In September 1970, the weather on the firing ranges at Fort Lewis.Washington, was in the 70's, with gentle breezes, the bluest skies I ever saw, and with a dramatic view of Mount Rainer. The orientation training for Vietnam included rifle firing, emergency medical care, cultural training, and other items I can't remember.

One interesting briefing was that the Vietnamese world view is different from the Western view. That many Vietnamese are Buddhist, and they believe that American soldiers are also deeply religious, because they consult their god many times a day for direction. It is common when talking with Americans that they will glance at their god for advice and suggestions. Their god is very personal and always close by. They wear their god usually on their left wrists. That god is a watch that American soldiers wear. We thought it was quite funny. Yet, it did make a point, they have a different view.

At the end of the training, a clerk handed me an airplane ticket to Vietnam along with promotion orders to Captain. Immediately, I went to the base exchange and bought the Captain rank insignia. Then went to the Officers Club and bought a round of drinks to celebrate with strangers. Quickly, I said goodbye and headed for the plane.

But the plane was delayed which allowed me time to explore a bookstore in the terminal. A couple of guys looking for a fight started pushing and shoving me. But I just looked at them, shrugged my shoulders, and moved away. I thought, "Don't do something stupid."

I walked to the airplane gate and boarded the plane.

Flight to Vietnam

While the plane refueled in Alaska, we were not allowed to get off. However, in Japan, while the plane was refueling, we were allowed to walk inside a small courtyard with razor wire on top of the walls. It was strange to see Military Police with weapons guarding the area. At first I thought they were protecting us from hostile Japanese. But then I realized that they weren't looking outside of the compound wall; instead, they were facing us with their rifles pointed at us. Then I realized, they were making sure that no one would desert to avoid going to Vietnam.

Getting off the airplane in Cam Ranh, Vietnam, the air was so hot and smelled it was as if I'd been hit in the face by a freshly filled wet baby's diaper.

I boarded an Army green bus. As the bus drove off the airfield I noticed that there were bars on the windows and thought, "Why are there bars on the windows?" As the bus went along barren sandy dunes, I thought, "When do I get a weapon? How am I supposed to fight an ambush when there are bars on the windows and no weapon?" Later found out that the base was so large that we had not even left the base.

Reception Center

When I got off the bus near the Reception Center, I needed to use the men's room. It was a vast hall with perhaps a hundred cubicles for individual toilets. Many women were bathing children in the toilets. They washed the child. Flushed the toilet. Washed the child. Flushed the toilet. As I passed the stalls there were more women washing their children.. I lost the urgent desire to use the men's room at that moment.

However, I was puzzled about why there were so many Vietnamese women on the base and particularly washing their children in the men's rest room. It turned out that the US Army hired Vietnamese men and women to do various chores on the bases such as working in the barbershop, washing clothes, cleaning rooms and facilities, and cleaning both the flush toilets, and the various sizes of outhouses. So, apparently those Vietnames women that I first saw were permitted to bring their children onto the base.

At the in-processing Reception Center, I filled out some forms and received an instruction sheet about where to eat and sleep. Eventually I found a bunk and was getting ready to sleep. An officer claimed a bunk next to me. He had Airborne, Ranger, and Special Forces tabs on his uniform. Quietly he made it known how great and strong he was. With practiced fanfare, he carefully placed his rifle underneath the mattress. Then laid on the bunk bed and went to sleep. I wondered for a moment how did he get a rifle so quickly.

Later, that night, as I was comfortably falling asleep in my bunk, when I heard a buzzing sound like an overloaded transformer about to explode.

Figure 2-1 An example of a helicopter shooting a minigun

Except it was intermittent. The loud monotone buzz lasted a few seconds. Then quiet. Then buzzed again. Over and over this happened. "What the heck?" I got up, went out to see what was happening and wondered why no one was repairing the transformer. Everyone else remained asleep as though this was normal, or they didn't care.

Looking up at the velvety blackness, I saw that when the buzzing sound came on, a long red rod appeared from the sky going down to the earth. When the buzzing sound stopped, the rod disappeared. Duh, it was a helicopter shooting a minigun at some target. It could shoot up to 4,000 rounds per minute. Every five or seven rounds there was a tracer round which made a red streak. With so many rounds a minute being fired it made the appearance that the helicopter had a long broomstick sweeping an area.[1]

Welcome to Vietnam

In the morning during in-processing, it seemed as if the first and most important thing was to starting counting the days till going home. However, I did not do that. A former Command Sergeant Major counseled several of us to not count the days because that just made the time feel like a prison and would seem forever.

---

[1] Copyright © International Ammunition Association. Photo obtained by and Forum Maintained by Aaron Newcomer: https://forum.cartridgecollectors.org/t/type-of-7-62x51-ammo-used/46593 November 10, 2022.

Nevertheless, many soldiers formed calendars where they counted down the days until it was time to go home. Sadly, it made the soldiers more interested in getting through the days to go home as compared to winning a war. Their only interest in winning was getting on the Freedom Bird, on which they would leave. No one cared. There was no enthusiasm to win the war; instead, it was put in the time and then go home.

I stood in line to get my assignment. The same Airborne Ranger from the previous night stood in front of me. At the assignment desk, the lieutenant looked at the Airborne Ranger's qualifications and handed him an assignment. The guy went ballistic. "You can't assign me to that unit."

"Sir you meet the exact qualifications for Special Forces."

"Look I want to go to the 101st.".

"Sir! You are going to the Special Forces unit in dumb fuck who knows where. The Army paid a lot of money to train you. Those Airborne, Ranger, and Special Forces tabs aren't for show. YOU are needed *now* in that unit!!"

The guy's composure changed. He angrily slunk away.

Then, I stepped forward. The Lieutenant (LT) glanced at my qualifications. He grabbed his engineer officer pile. Pulled out the top vacancy paper and handed it to me.

Remembering the previous exchange between the LT and the Airborne Range, I said, "Yes, I am qualified for that unit. Perhaps you have a unit in which I have already trained for two years. In that type of unit I know the equipment, the capabilities, and how to employ them. I can step right inand do the job. Please, look in the pile of engineer assignment vacancy papers and see if there is a unit that I am fully trained in? There must be some vacancies that are an exact match for the Army and everyone wins."

He sighed, perhaps feeling that this was going to be another argument first thing in the morning and what a lousy day it would be. He shuffled a few papers in the pile and handed me an exact match. I was ecstatic. Floated out of the tent into the cool embrace of 100 plus moist degrees. I didn't even notice the odor from the latrines.

To the Engineer Brigade…

Another drab green Army bus dropped me off at the 18th Engineer Brigade HQ in the afternoon. A captain greeted me, scheduled a time to meet the Commanding General, and arranged a room for the night. He said I would go to the group headquarters in the morning by helicopter.

The Commander, Brigadier General Henry C. Schrader, was enthusiastic, tired, and very personable. He explained a few things, inquired about my background, looked at my file and said, "I am glad that we are getting experienced Captains now. Good luck." The General made the comment about experience because there were so few captains. Most of the soldiers that were eligible after two years to become captain opted to get out of the Army.

In the morning a helicopter brought me…

To the Engineer Group…

A C130 transport plane took me to an airport near the Engineer Group Headquarters. After arriving, I filled out some papers and was assigned a bunk to sleep. In the morning someone told me to be in charge of a convoy. That was truly puzzling. I didn't know where it was going, what it was capable of, what the communication codes were, who, how many people, what trucks, and heaven knows what else.

The guy said, "Don't worry about anything. Nothing will happen. It is safe. Everyone knows what they are doing."

So, I hopped into the jeep and for a few minutes rapidly read various papers which he had given me. He then said, "Good luck" and walked away.

The convoy started up without my direction and proceeded on its way. I thought, "This is stupid. And dangerous for everyone. I personally did not instruct the drivers and the NCO's at a meeting on what they were to do in case of an ambush, validate the communication system, the time till the destination, the proper distance between vehicles, the speed, etc. etc. Welcome to Vietnam."

After calming myself, I enjoyed the sights. The women in their traditional dress. Men working in various shops. Gasoline was sold in various types of glass liter bottles. I assumed that the bottles stayed at the gas station. The sky was blue and weather was quite pleasant. The convoy passed through several villages.

As it left one village, there was an old lady dressed in solid black, sitting on the right side of the road. Her face lacked emotion. It was stoic. No expression. Looking into her eyes was like staring into a deep, dark abyss. Fifty-three years later, I can still see those dark, emotionless eyes. Unconsciously, I waved a V for peace with two fingers. Without emotion her eyes followed me. The convoy moved on.

The next day I was assigned to run the same convoy. I thought, "Is this what a newly promoted Captain is supposed to do in Vietnam?" However, this time I had everything better organized and I gave the order for the convoy to move.

Again, the old lady dressed in black was by the roadside. Again, I waved the V for the peace symbol at her. There was absolutely no expression on her face. Her emotionless eyes seemed to be a portal to another universe. I felt being sucked into her world. I kept waving the "V" finger sign at her. The convoy kept going and I didn't look back at her.

This went on for several days. Then one morning as I was getting the convoy ready, a clerk ran out of the Colonel's office. "Sir! The Colonel wants to see you now."

"Who is going to take the convoy?" I asked.

"Don't worry about it. Another guy will take it."

I shrugged. Got out of the jeep. Went in to see the Colonel.

"Captain, here is your assignment. We might get you out today. But I think the vehicle has already left."

So, I had a day off. That evening, my roommate came in. "You know that convoy that you were running? The guy who took your place was killed. I think it was that lady who you mentioned that blew him up." Whether that was the convoy I would have taken or some other convoy in some other unit I'll never knew.

Numbly I felt, "Glad it wasn't me." Gradually, I was becoming numb.

Welcome to Vietnam."

Chapter 3
Intelligence Officer

The work that my Battalion was doing was supporting the fighting shown in the three pictures below; but, none of us were involved in direct combat. The bodies were always of North Vietnamese or Viet Cong soldiers. I never saw any American, Korean, nor Australian soldiers dead from battles. While traveling on various assignments, I occasionally came upon scenes like this after a fight. So, I had it good. Was comfortable. Had regular food, shelter, and somewhat decent sleep. The war movies are about blood and guts scenes shown above. About heroes, anger, killing, maiming, bombing, losing, and hoping to win. That was not my war experience.

Hopefully these pictures will help readers understand some of the context of rest of this story.

Figure 3-1 Engineers built roads, bridges, and airfields in support of these operations

Figure 3-2 A typical result of combat operations which the 84th Eng did not directly support

Figure 3-3 Seeing the dead along some roads or fields was a frequent occurrence.

Assignment

Somehow, I arrived at the 84th Engineer Battalion Headquarters in Quy Nhon. Lieutenant Colonel Remus, the Combat Heavy Engineer Battalion Commander greeted me. He was a confident, sharp, inspiring, coach, and commander with whom I stayed in touch for the next fifty-three years. His leadership and coaching helped me in both my military and civilian life. Truly an inspiring person. A mensch. Very personable and patient with each everyone while still pushing to get the job done. He was always there with us. Never once did he hang me out on to dry on a limb when something didn't go right. Thus, he led, trained, mentored, and got the job done. He should have been a general officer.

Figure 3-4 Cam Rahn to Quy Nhon

Boring

I was assigned as the Battalion Intelligence Officer, which in an engineer unit is very boring. Every morning reading situation reports about a North Vietnamese or Viet Cong unit 'asking' for bags of rice in some village, or attacking some convoy, or blowing up a building, or, or, or. It seemed like useless information for engineers. Though, I quickly realized that the key was to keep people aware that though the Battalion was not fighting, there was still a war on. So, everyone had to have some understanding that there was still a threat and not become lackadaisical. Thus they needed to follow procedures to be safe while doing the engineering work.

While it was a disappointment to have this assignment, I decided to thoroughly and enthusiastically do the job while volunteering for other projects as needed.

Battalion Security

One of my other responsibilities was base security of the Battalion Headquarters and the co-located Company C (One of the five subordinate engineer units to the Battalion. There was a perimeter of razor wire and guard towers. Each night a different officer checked the perimeter every two hours to make sure it was intact, climbed each tower to stimulate the soldiers to stay awake, checked the communications, and then slept for an hour. Throughout the night, he repeated the process. In the morning he returned to his normal duties. It was a boring duty and easy for soldiers on guard duty to fall asleep. There were motion detectors which seemed to detect only animal movements. Those created false positives which became easy to ignore. If there was going to be an attack, then more motion detectors would indicate that something was going on. Also, since we were engineers many of us thought that there wasn't much reason to attack the compound.

Additionally, many of us heard that the North Vietnamese and the Viet Cong believed that they would win the war. So, the work we engineers were doing would benefit them after the war was over. It would be against their best interests if they attacked an engineering unit. Thus, they wouldn't attack us engineers.

One night, as I dozed off in my comfy bunk, it felt as though someone turned on a huge aircraft search light into my eyes. Less than a fraction of a second later, the volcanic sound of the ammunition depot blowing up across the street fully woke me up. I thought, "I screwed up!"

My butt suddenly expanded so wide that the Titanic and the iceberg it hit could have passed through. In the next fraction of a second it tightened up so much that an electron whirling around the nucleus of an atom could not pass through. For the first time time I understood the GI jargon about the pucker factor.

Everyone woke up without the alarm needing to go off. They grabbed their rifles, and ran to their stations. A piece of shrapnel slightly cut my right foot. I did not want a medic. But one came anyway. SGT Ed MacNeil III seemed disappointed that it was not more serious. I felt embarrassed and sorry for him because he wanted to really help someone that needed it. Also, he was a super conscious, dedicated, and well liked. Sadly, several months later a mine killed him and four other soldiers.

Later we went back to sleep. Got up in the morning. Then realized that the 105 and 155 artillery ammunition was still exploding and continued to do so for two days. As best as we could we continued operations. I never did find out what actually caused the blow up. Perhaps someone smoking, a rocket, or sabotage?

Monsoon

During the monsoon heavy rain flooded the headquarters. Still, we kept at our jobs.

A hillside by the compound started to slowly slide down towards the HQ and Company C compound. So, I cobbled together a crew of soldiers, steel sheeting, poles, and sledgehammers. We went up the hill and built hasty barriers to prevent more mud from sliding down. While it was wet and dirty work everyone made it a fun job. Perhaps we were just like children playing in the mud again. Being outside in the fresh air was a welcome change from the unrelenting tedium of office paperwork, and we were just like children playing in the mud. Great fun!

For several days the Battalion headquarters was knee deep in water. As best as we could, we continued doing our jobs. To prevent infection if anyone was cut by objects under the water, Lieutenant Colonel Remus mandated that the medics give everyone tetanus shots. Considering what the grunts were going through, no one complained because we had it good.

Bong Son Bridge

Before, during, and after the monsoon part of my duties was visiting various work sites and making reports about the quality of the culverts, bridges, and

Figure 3-5 Bong Son Bridge under construction by Company B, 84th Eng Bn (Combat Heavy) Courtesy of the 84th Eng Bn Association Vietnam.

road work. One of the projects was to check the quality of a steel girder and concrete bridge about 1,000 feet long. It was about seventy-five feet high, and thirty-feet-wide and across a three-foot deep. The last part of the project was to install an angle iron railing ("L" shaped steel) as shown in the picture on both sides of the bridge.

My driver and I stayed at night in the nearby 173rd Infantry Brigade compound. Meanwhile, during the night, a South Vietnamese infantry platoon guarded the bridge. One morning my driver and I left the Brigade compound. As we drove through the village, I was pondering various issues and what to check. Suddenly, my driver exclaimed. "Man. Look at all the angle iron!" Interrupting my deep thoughts, I became dimly aware the little restaurant had angle iron. The gas station with the liter bottles filled with gasoline had angle iron. The whorehouse had angle iron. It seemed as though every business in the village had angle iron. Suddenly, it occurred to me. "Where did all this angle iron come from?"

When I arrived at the bridge, the platoon leader and his platoon were gone. It rapidly became clear that the platoon leader had sold the angle iron, paid his platoon, and disappeared. Since there was a lot of interest in getting the bridge finished, the word rapidly spread among important Vietnamese and American people. Arguments started as to who was responsible. After dinner, there was a heated meeting. The Army advisors were yelling at the Vietnamese politicians, who were yelling back and at other Americans. Suddenly, they all looked at me and started yelling that I was responsible and would have to pay for all the angle iron and that I had to get angle iron now.

"What? How did I get mixed up in this?"

Stunned and depressed, I walked out of the tempest. For a moment I considered how many months of payments I would have to make for the loss of the angle iron. Then I looked up at the black sky. The stars were like quiet jewels shimmering. I thought of the millions of other soldiers through the centuries that also must have looked up at the same stars with the same depressed feelings of much more serious calamities like death, maiming, battle loss, and impending doom. In comparison, my problem was tiny and insignificant. In fact, in the grand scope of the universe my problem wouldn't be even register. So, I just needed to figure out a way to solve it. And besides, regardless of what the powers to be thought, there was no way that I was responsible for either the security of the bridge or the loss of the angle iron.

It became another learning experience. When something goes wrong, people run for cover, hope no one blames them, and to defend themselves they immediately pass the blame onto someone else. They'll blame anyone hoping the problem doesn't stick to them. The aim is to not get stuck with the blame. This was a pattern I would see over and over in both the Army, civilian life, politics, seminary, religion, etc. In fact, it is one of the first stories in the Bible. "Am I my brother's keeper?"

Somehow in a few days more angle iron appeared. I never learned where it came from nor who resolved the issue. No matter, I was not blamed nor had to pay for anything. Perhaps the Military Police went through the village and confiscated all the angle iron. I really don't know and decided it was more prudent to not ask

Why Such a Long and High Bridge?

For several months I wondered why engineers had designed the bridge to be so long and so high. As the project was finishing the monsoon season was starting. The rains at times were so heavy that it was impossible to see more than thirty feet away. As the water level rose the thirty foot wide stream became a 1,000 foot wide river. Then I realized that in the far distance there were mountains. All that water in those mountains flowed under the bridge.

After a few days, I wondered if the bridge was high enough. Would there be room enough for the dead water buffalo and people to float by? Somewhere in the rear echelons of the Army, some design engineer precisely figured out the water flows and the requirements for the bridge. Ever since then I have wanted to meet the designer and learn how he did it. Conceptually, I have the idea; but the engineer in me wants to know the details, the calculations, the politics of building it, etc.

Also, all that water flowed into the nearby village on one side of the river near the Company B compound. I often chatted with those villagers. As the river rose and crept towards the village, I mentioned that their huts would get flooded. Perhaps they should build their huts higher up away from the river.

They laughed and said, "Captain, you crazy. If we build further up, then we walk further to use the river. When the water comes into the huts, we sit on top of the roofs. We only sit on there for two weeks. The rest of the year we are close to the river and don't have to walk so far." They laughed at my silly suggestion.

It was another cultural perspective. We all look at the same events and frequently make different interpretations. Is there a right or wrong interpretation in this case?

Battalion Moves to Da Nang

The Battalion was ordered to move from Quy Nhon 200 miles north to Da Nang. The Battalion S-4, (Logistics Officer), needed help with the move. So, I was (and still am) a free-floating spirit, Lieutenant Colonel Remus assigned me to help the S-4 with moving the equipment by sea.

At the dock there was a ship to take the equipment and a Warrant Officer controlled the dock operations. He needed the TCMD's (Transportation Control Movement Documents) exactly filled out in multiple copies. He was quite heated about it, since apparently none of the documents were precisely completed. So, he patiently explained that the load masters and crews use the documents to position cargo on the ships. They look quickly at very specific places on the documents for information. Those spaces were blank. And if they start hunting for it, then they might make look in the wrong spot and use the incorrect data. That can cause problems with loading, and if the sea is rough, improperly loaded cargo can cause the ship to list and even take on water. Everything had to be so precisely correct and in the right position on the form.

I thanked him for patiently explaining the necessity of precision filling out the documents. This was another lesson in making sure that people need to know the importance of doing quality on-time work. If they don't understand that, then they drift their own ways and mistakes are made. In the military those mistakes even by support operations can get people hurt or killed.

I quickly figured out that to go back to the Battalion there would be a fight to get some clerks to properly fill out the forms. Furthermore, none of the documents would be started until the following day. So, I asked the Warrant Officer for a typewriter, forms, and a place to work. Within a few hours, the forms were completed to his satisfaction and then equipment could be accepted for loading.

Four Star South Korean General

Some days later, the equipment was being loaded onto a freighter. To make sure that everything was going smoothly, I went to the shipyard. As my jeep approached the entrance, a South Korean MP was directing traffic. Smartly dressed in a crisp uniform, with white gloves, and shiny helmet, he pointed me towards the dock shaped like a "T." My driver turned the jeep as directed onto the base of the "T." The ship had docked at the far left leg of the "T."

On both sides of the long pier there were rows of smartly dressed South Korean soldiers, standing at attention, wearing white gloves, rifles at their sides. They were perfectly spaced, six ranks deep, and at least a football field long. My driver jokingly said, "Man look at this. They are all here for us."

As we neared the junction of the "T," a perfectly dressed Korean MP saw us. With great pride, he used his right arm mechanically to wave us on. In astonishment, he looked across the intersection and with his left arm began mechanically waving another vehicle through it.

My driver jammed the brakes, stopping the jeep inches from T-boning the General's jeep. It had a bright red rectangle with four huge white stars on the front . Sitting up high in the back was a four star South Korean General. With his arms folded, he looked down on the MP, growled with an angry face. His jeep continued on. Immediately a crowd of screaming

South Korean soldiers rushed at the MP. A Lieutenant began slapping – not slugging – the MP. The crowd was screaming at the soldier.

I calmly, yet with a sense of urgency, said to the driver, "Get the fuck out of here – however, slowly – before they turn on us." Passing through the crowd of soldiers who meekly let us by, we continued to the ship.

After verifying that the cargo, paperwork, and manifest was all in order and ready to leave, I returned to the jeep. Satisfied with the loading of the ship, my driver and I headed back towards to the intersection. The MP and Lieutenant were alone. The Lieutenant continued to slap the red face of the teary MP.

My driver hesitated. I motioned for him to drive on. I was not about to get involved in some sort of inter-army incident which I would be explaining for the rest of my life. Nor did I want some other nearby Korean soldiers, working on various tasks, to see me involved with the MP and deciding to see what would be happening. And then suddenly my driver and I would be involved in a fight. A fight that I would not win.. We left the Lieutenant slapping the red faced crying soldier..

I wondered many times what ever became of them.

Chapter 4
Company Commander

While in the position of Intelligence Officer I asked my Battalion Commander, Lieutenant Colonel Remus, to assign me as a Company Commander. Some days or weeks later he informed me that I was to be the Company B Commander, which delighted me. Since I was familiar with the unit mission, equipment, issues, and the leaders, it was a perfect choice. Then about a week before I was to assume command, the Group Commander wanted someone else who was outside of the Battalion to be in charge. So, the Company B command assignment was out. That decision saved my life. That story will be told in a later chapter. Strange how some decisions happen that have a dramatic effect on one's life without knowing it.

About a month later, Lieutenant Colonel Remus told me to assume command of Company D. He furthered explained that Company D of the 84th Battalion was being transferred out of the Battalion and Company C of the 589th Engineer Battalion was replacing it. It would then become Company D of the 84th Engineer Battalion. That I would immediately move it from its location 250 miles south of the 84th Engineer Battalion Headquarters to 450 miles north by land, sea, and air. Meanwhile the Battalion was also moving north from Quy Nohn.

Assumption of Command

I flew by C130 Air Force plane to Phan Rhang Air Force Base. The last leg of the flight was by a Huey Helicopter west to the Company camp site at Song Pha. The helicopter had the standard M60 machine guns and gunners on each side. There was another passenger also assigned to the Company. Because of the noise there was no way to have a conversation with the new person and I never got to know him.

The Company area was on a flat plain at the base of a mountain range. The compound had a perimeter of razor wire and guard towers. Shortly after meeting people, learning and getting to understand the projects, and the area, there was a change of command ceremony. The First Sergeant called the company into formation. He took the flag from Captain Kropp, the Commanding Officer and passed it to me. It was at that point I assumed responsibility for all that happened and didn't happen in the unit. He briefed me on the unit, making my head swirl in information. A day or two later he left. Years later he was promoted to Major General.

The unit was authorized 160 soldiers and three lieutenants; but instead, there were two lieutenants, Don Scholtz and Ken Ament. They were superb, thoughtful, caring, and had excellent judgment. Similarly, the Chief Warrant (CWO 2) maintenance officer was quietly forceful, knew what he was doing, and highly dedicated to the mission and his men. Apologies to him and to my First Sergeant for not remembering their names. We called them Chief and Top respectively. Chief in honor of his being an expert in maintenance which kept the unit going, and Top in honor of his being the senior ranking Non-Commissioned Officer (NCO) in the unit. We were fortunate to have such capable people. While it hurt not having the full complement of lieutenants, one takes what one has and then makes the best of it. The lack of resources is no excuse for not getting the work done.

The main priority was security, finishing the projects, signing for and taking responsibility for all the equipment, and starting to move the company. The equipment consisted of dozers, graders, scrapers, dump trucks, support vehicles, maintenance equipment, rifles, machine guns, plus unauthorized equipment like four 50 caliber machine guns, trailers, spare parts, and other items. Without those 'extra' parts, the mission would have been difficult to carry out.

Court Martial?

A day after assuming command, a convoy was leaving the compound for a project site. The standard procedure for security was a minimum of two vehicles leaving the convoy at a time. That way if something happened to one vehicle, then there was help with the other vehicle. One of the soldiers in the unit made friends with the new soldier who was on the helicopter with me. They wanted to find some women for an afternoon of pleasure instead of working. So, they borrowed a vehicle. Followed the convoy out the gate and turned left towards fun while the convoy turned right towards work. Further down the road, they were ambushed. The new person was killed. Bullets scraped the front and back of the other soldier's neck.

The new person was so new in the unit, there was some difficulty of firmly and positively identifying his body. Eventually, I was satisfied that this had been successfully accomplished.

His injured companion was taken to the Air Force Hospital at Phang Rang about an hour away and I went to visit him. He was in a ward with twenty or more beds. I was uncharacteristically angry about what he had done. So, I told him that I should have him court marshalled for not following orders and procedures, misuse of government property, being absent without leave, dereliction of duty, and involuntary homicide (or some such similar term).

However, I didn't have time to start and follow through with the complexity of such a court martial. The Company was still in the 589th Engineer Battalion and I was still in the 84th Engineer Battalion. The issues would have been which Battalion was going to do the legal work. The 589th was going back the US. The 84th was moving. I was short of officers, capable NCOs, and soldiers. This was a burden that would have reduced the effectiveness of the mission.

So, I told him, I hoped that he remembers that he killed someone. That he would live with that thought for the rest of his life. And that he would make something useful of his life. Turned around and walked out.

Did the soldier breathe a sigh of relief? Did he care? What ever happened to him? Did he later kill himself, become a bum, or perhaps turn his life around? I don't know and still wonder.

Back at the company, I wrote a condolence letter to the mother of the deceased soldier. Dear Mrs So and So (I had the name and now sadly don't remember). It went something like this: 'Your son was horny. So, he, without authorization, took a vehicle, failed to follow procedures and orders by leaving the compound and headed towards fun with some whores. On the way he was killed in an ambush.. His friend was wounded on the front and back of his neck.'

Later I tore it up into little pieces and wrote the standard "I regret to inform you...." Whatever happened to the letter I don't know. All I know is that his death did not count on the 84th Engineer Battalion list of casualties since the unit was not yet in the 84th when the incident happened.

On and off since then I have pondered if this was this the right way to have handled the situation. I don't think so. Was this dereliction of duty on my part? Perhaps. Yet even today, I would have done the same thing. There was just too much going on and the chaos of a major court martial would have caused too much disruption to the workload of my two lieutenants, the First Sergeant, some of the Non-Commisoned Officers, let alone calling back the former commander Captain Kropp and who knows what other officers in the 589th that was moving back to the US. It was a decision I made to focus on what was in the best interest of successfuly completing the mission So, the information was passed up to the Battalion that still controlled the Company and I let them decide what to do.

Drugs and Race Relations

As with so much in life, one has plans and goals to do so many things, but life happens and one gets consumed by other events. My plan was to focus on completing engineering projects, and moving the company by land, sea, and air.

Instead, the deterioating support for the war showed up in the unit. Eighty percent of the soldiers were drafted. They did not want to be in Vietnam, let alone to be in the unit. Their country did not support the effort. They were tired and bewildered by what was happening. The career Non-Commissioned Officers were tired and had served several one year tours of duty in Vietnam. Everyone just wanted to go home. .

I was accountable for accomplishment of the mission, maintenance of all the equipment, percentage of spare parts on-hand compared to what was authorized, re-enlistment rate, percentage of assigned soldiers compared to authorized, discipline, Article 15 rate (formal company commander discipline), and the venereal disease (VD) rate. I thought, "The VD rate? You have to be kidding. I am not inspecting soldiers' private parts for that." As a teenager I heard soldiers from WWII and the Korean War talking about 'short arm' inspections to keep the VD rate down. I never knew if that was true; but, true or not, I wasn't going to do such inspections. Although, I was responsible for everything that happened and didn't happen in the unit. I still wasn't going that.

Each morning before breakfast I inspected the compound. There were small plastic vials scattered around. What were those things? Yet, prior to the morning formations in the compound, Top had the area cleaned up and the vials disappeared. Someone explained to me that the vials had heroin. It was then that I realized why the soldiers looked so haggard. It wasn't only from being overworked, it was from being high on heroin or marijuana. So, each morning I counted empty vials without telling anyone. Roughly, half of the one-hundred and sixty soldiers were using some sort of drug. Then I noticed that about a quarter of the others had hangovers from drinking too much.

On top of that, the racial tensions in the US were clearly visible in the unit. The Southern Whites hated the African Americans. The African Americans were angry at the Whites. For a while there was nothing serious, just a lot of obvious tensions. Tensions that would eventually become serious.

The movie, "The Ten Commandments" with Charlton Heston playing Moses came to mind. The Pharoah played by Yul Brynner was highly frustrated with the Israelites and Moses. So to reduce the problems, the Pharoah said, "Work them more."

So, I thought these guys have too much time on their hands. I have got to find much more work for them. That way all they will think about is sleeping, eating, working, and getting laid. They won't have time for race problems and taking drugs. They'll be too tired. Over and over that proved true. When they had time on their hands, some got into trouble.

What a strange way to conduct a war. I began to realize that the news media, movies, and television influence people's actions and their meaning of life.

Can We Get Some Entertainment?

A couple of the NCO's asked me if they could get some entertainment for the soldiers instead of movies which were projected against a bedsheet or a white board draped on the side of a building. They suggested that it would be good for the morale, be fun, and they neededsome live entertainment. I thought that it was a great idea and told them to go ahead.

The following evening the NCO's brought ten to fifteen Filipino women to the show. The women sang, danced, and showed more than I expected. The soldiers were delighted, applauding, and laughing. It was the first time that instead of being somber, tired, and dragging their bodies around, that they had showed so much enthusiasm, joy, and happiness.

The next morning. The African Americans and Whites were high fiving each other. Happily greeting and chatting. Wow, I thought. Nice.

Then five angry soldiers rushed me. They were incensed at what had happened the previous night because the women had sexually satisfied many soldiers. That it was immoral and against God's will. They demanded that I should do something about that.

Taken a bit back, I was puzzled for a moment. Then said to the five fundamentalist Christians, "Did you do anything with the women last night?"

They said, "No! Of course not. We demand that you do something."

"Well look around. Everybody is happy. You guys are the only ones that are angry and you are pure by resisting temptation. God will be happy with you."

They responded, "We'll tell the general."

"That's your right." Inwardly I sighed. Another problem to handle.

They turned around and marched off in a huff muttering and … I never heard anything more.

I felt like the time when some people told Abraham Lincoln that General Grant was a drunk. The President responded that he fights and wins battles. So, find out what he drinks and send him a case of it from me. I thought, "Mmmm…maybe the soldiers need more shows." But no, I didn't follow through with Lincoln's advice to repeat such shows.

However, for the remainder of my time as the Company Commander, I had informal discussions with many of the soldiers of the risks of sex with prostitutes. The discussions were friendly. They usually said something like, "Sir, if we get something we'll just go to the Doc (the medics assigned to the unit) and get the shots. There's nothing to worry about." Maybe the discussions helped. Maybe they didn't.

Months later after the move from Song Pha to Phu Bai, the Battalion surgeon visited the unit checking health and welfare issues. Upon finishing he reported to me that the VD rate was satisfactory. I was pleased that I wouldn't get my hand slapped for that.

These events helped me to understand that the sexual drive and the need for the comfort of a tender and loving touch from someone is fundamental to a meaningful life. People will satisfy those needs in ways that others will smugly condemn, or overlook. Some will quietly pursue in their own ways to fullfill their desires and manage to get societal approval, while approve of killing and maiming the enemy.

Preparation to Move

As the Company was completing construction projects it also prepared to move from Song Pha to the Air Force Base at Phang Rang and then north to Phu Bai.

So, I coordinated with local Vietnamese Army officers to turn over the compound. It was part of the overall drawdown of US Forces to pull out of Vietnam and have the South Vietnamese Army fight the War.

We inventoried the buildings, the guard towers, and then walked the perimeter of razor wire. These discussions were supposed to be secret, known only to some Vietnamese officers, and my unit. In the morning as the trucks were lined up ready to pull out, the local population gathered around the perimeter. Meanwhile, the Vietnamese officers and I quickly reviewed the compound and signed the documents transferring the compound to the Vietnamese Army.

Just as we were ready to pull out, my jeep had a mechanical problem. Some guys quickly suggested putting the jeep on one of the lowboy trailers which was pulled by a tractor. It made quite a sight for me to ride up high like a conquering Roman General hero or perhaps it looked just ludicrous. I don't remember if I rode that way or went in one of the jeeps. I think it was in one of the vehicles.

As the long convoy pulled out, the local Vietnamese villagers climbed over and through the perimeter. They began to pull light fixtures, boards, and anything else they could carry out of the buildings. One lone Vietnamese officer was running around trying to stop the scavengers.

I pondered the situation for a few seconds and decided that it was his problem, not mine. That if I stopped to help him, I would be dealing with a convoy open to being pilfered and fights between the soldiers and the Vietnamese civilians. So, I decided to keep going. Never did find out what happened to the compound, nor to the Vietnamese officer. Another mystery.

The convoy arrived at the temporary facilities in Phan Rang. A week later, I went to Phu Bai to verify that the quarters, ration ordering, maintenance area, parking area for the graders, pans, dozers, trucks, jeeps, tractors and trailers, generators, miscellaneous equipment, etc. was adequate. Also, I checked that the project load with the required materials was available. In the meantime, the two outstanding Lieutenants and the Warrant Officer moved the Company some four-hundred and fifty miles to the north.

1LT Ament suggested that he be temporary Company Commander when moving over one hundred soldiers to the LST (Landing Ship Tank), while on the LST, and then to Phu Bai.. The suggestion made sense in the event he had to deal with disciplinary problems. Which fortunately he handled without the need for Article 15's. Here is his story of transferring Company C, 589th Engineer Battalion to become Company D, 84th Engineer Battalion and then moving the Company from Phang Rang to Phu Bai.

Chapter 5
Moving the Company
By 1 LT Ken Ament

Around February of 1971, rumors started flying that our unit (the 589th Construction Engineer Battalion) was going home. Charley Company completed the road projects from Phan Rang to Dalat (QL 13—now QL20), and the work that the Battalion was doing down south on QL1 was also winding up. We at Charley Company had a surprisingly good gig. Despite the occasional mine or ambush on our road, we were fifty miles away from Battalion headquarters (and all its formality) in one of the most beautiful settings one could imagine. Our unit was based at the foot of the Ngoan Muc pass on the Da Nhim River in the quaint little village of Song Pha. The road we built ran fifty miles from Phan Rang to Song Pha, then a 30 mile stretch up that pass, a 9% grade, to Dalat. The Ngoan Muc pass was one of the major engineering accomplishments of the American war effort in Vietnam. So, we were ready to go home.

Figure 5-1 Ngoan-Muc Pass *Figure 5-2 Culvert Construction near Song Pha*

I had served nine of the required twelve months tour of duty. So, I would probably get an early out when the 589th stood down and I began making plans for a trip to Europe, then graduate school. Not so fast, Lieutenant!!!! You're going to Phu Bai!!! President Nixon needs you there. Captain Kropp, you've got 11 months in country, you can go home (Tony Kropp went home, stayed in the reserves, and ultimately made major general). Ivan Beggs came down to Song Pha from the 84th Engineer Battalion, our new unit, met everyone and then headed for Phu Bai with an advance party to prepare our new home.

So 1LT Don Schlotz, Tom Brennan, Bill Krikorian and I (four OBV-2s (Obligated or Volunteer with 2 year commitment) with questionable attitudes) started preparing to move the company north. Since I had seniority, I was put in charge, even though I was the youngest. The rest of the 589th started making their plans to go home (lucky bastards). A major from Group showed up one day with official secret orders that told us that we would head north on a certain day. Gee thanks, major. Everyone in the village knows that, but thanks nevertheless.

Figure 5-3 1LT Ament leaving Song Pha

One of the largest problems we had during that period was keeping the troops occupied. Too much free time and not enough work allowed drugs and racial issues to come to the forefront. Finally, three weeks later, we headed out of Song Pha for the last time and drove all of our equipment and belongings to Phan Rang, where we were to meet up with two Korean LSTs to take us north.

The map next page shows the area of operation west of Phan Rang across a flat plain, through the mountains into Da Lat. The company then moved north to Phu Bai by LST. It was about 50 miles north of the Battalion Headquarters in Da Nang.

We were to stay in Phan Rang overnight and load the LST boats the next day. Unfortunately, the LSTs showed up a couple of days late, allowing

Chapter 5: Moving the Company 71

Figure 5-4 Arrows show the company movement from Song Pha and then by LST to Phu Bai.

our restless troops even more time for fighting and drugs. We were so concerned about the drugs on the boats that we got the MPs in Phan Rang to bring a couple German shepherd dogs out to the docks where we were boarding, told the troops that these were drug sniffing dogs (which they weren't), gave the troops a five minute amnesty period to get rid of their drugs in a couple of trash cans, and boarded the ships. There was a ton of marijuana and heroin thrown away.

Figure 5-5 Loading the LST

We spent three days on the LSTs, fighting seasickness and drug withdrawals by some of the troops. We pulled into harbor about 30 miles from Phu Bai around 4:00 p.m. in the afternoon and convoyed our unit and all its equipment to our new home in Phu Bai. We got to the barracks around 10 that night. The next day CPT Ivan Beggs took the unit back over, and I was delighted to go back to my easier role as earthwork platoon leader.

Figure 5-6 On the LST from Phan Rang to Phu Bai

Figure 5-7 On the road from the dock to Phu Bai

Figure 5-8 1LT Ken Ament is pondering the current situation while leading the convoy. The ever present children like to get near the vehicles and help themselves to 'souvenirs'

Combat Heavy Engineer Company Organization

By this time you might be wondering what a combat heavy engineer company does and is. There are several types of engineer companies: bridge companies, equipment companies, combat companies that are in direct support of division combat operations, and combat heavy companies that build roads, bridges, airfields, buildings, and large size bases. We were the later type company.

When we arrived at Phu Bai we also picked up support and supervision of the rock quarry and an explosive, ordinance, and demolition squad (EOD) which brought the unit to 180 soldiers. The last two sections are not shown in the organizational chart.

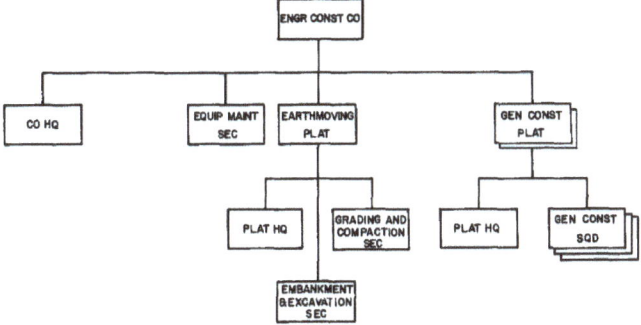

Figure 5-9 Combat Heavy Engineer Company Manning Structure

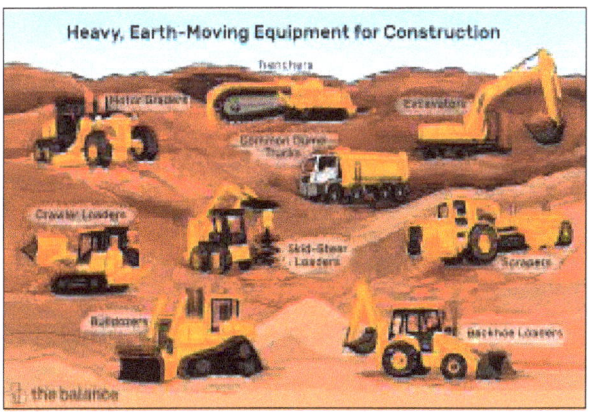

Figure 5-10 A civilian example of the Army equipment.

The equipment consisted of graders, bull dozers, front end loaders, earth scrapers, dump trucks, ten-ton tractors that pulled trailers, and miscellaneous support equipment. A civilian example of the prime equipment is shown in the picture.

Projects

Figure 5-11 Ken Ament, Red Dirt, Phu Bai, 1971

Figure 5-12 Trucks lined up to haul dirt for road repair and construction

Figure 5-13 Typical asphalt work

Chapter 5: Moving the Company 75

Time Off

Normally we worked six and a half days a week. Sunday afternoons were free time. Eventually, we worked just six days a week with Sundays off. Most of the guys slept or used some the recreational facilities of the 101st Airmobile Division (The Screaming Eagles). 1LT Schlotz and I took time to visit Hue.

Figure 5-14 1LT Ken Ament in Hue.

Figure 5-15 1LT Don Schlotz in Hue.

Chapter 6
The 101ˢᵗ Airmobile Div

This chapter describes general engineer support to the 101ˢᵗ Airmobile Division. The next chapter explains other related activities.

Coordination

Prior to the arrival of the Company, I coordinated with the 84ᵗʰ Combat Heavy Engineer Battalion, the 326 Engineer Battalion (Air Assault – organic to the 101ˢᵗ), the Division G3 (Division Operations) and G4 (Division Logistics). We agreed on project work, the unit location, responsibilities for security, logistics, etc. The only major issue was the company site location was way too small. After some discussions the G4 realized that the company footprint was substantially larger than their airmobile 326 Engineer Battalion Companies. We eventually agreed on a suitable site.

The 84ᵗʰ Engineer Battalion controlled which projects to work on with agreement with the 326 Engineer Battalion. Weekly I met with both Battalion commanders. For the most part, Company D completed the projects on time and to standard.

The jeep on the right with the CO's driver, Inge, is in front of the Company headquarters

Figure 6-1 The jeep on the right with the CO's driver, Inge, in front of the headquarters (Courtesy of 1LT Ken Ament)

Engineering Work

The projects were road maintenance, culvert repair and installation, airfield repair, erecting and repairing buildings, rock crushing, as well as explosive ordinance and demolitions (EOD). The Company normal troop strength was 160 engineers. The addition of the rock quarry and EOD detachment which increased the strength to 180 engineers which would normally have 4 Lieutenants assigned; instead we had 2 Lieutenants. The shortage of 2 Lieutenants created significantly greater workload for the 2 officers. On one occasion our project load required additional troops and so another platoon was assigned to us for two weeks. This brought our troop strength up to 220 engineers without the Lieutenant Platoon Leader. Another time one of our platoons was temporarily assigned to another unit, dropping our troop strength to 120 which felt like a vacation because there were fewer soldiers.

The constant variation in troop strength, projects, personnel rotating in from the US and out back home, venereal disease rate, plus the never-ending discipline, drug, and race problems made life interesting. It was nothing glamorous like in the movies. Just a lot of good men doing their jobs, not knowing why the country didn't support them, and wanting and waiting to go home. Many felt lonely for the companionship of their wives and girlfriends back home. There was an overwhelming feeling from most everyone of, "Does anybody care? Is anybody listening? When can I go back home?"

It was a credit to them that despite all these issues, the general context of the war, and the many arguments and discussions about the work, we did our jobs and then happily and quietly went home where no one cared.

Helicopter Flight Line

The most intense project was repairing the helicopter flight line. The monsoons washed much of the underlying sand away causing the expedient steel perforated mats to buckle. To remove the matting required cutting torches and a lot of heavy sledgehammer work. The soldiers took some pride for a while in that swinging the sledgehammer was similar to the "Big Bad John" song. Plus, they were getting stronger.

After removing the bent matting, they created a flat surface and installed new matting. The picture[1] is from WWII. Notice the perforated matting. The planes are WWII not the ones in Vietnam.

A large part of the Company was involved in the work. I briefed them on the project. Searching for a way to keep the unit from slacking off, I uncharacteristically said, "The work would be hard. *Our weaker sisters will not be able to keep up.*" It was not something I liked to say nor approved of. Yet it struck the nerve that I was searching for. Not something that today would be acceptable.

Figure 6-2 The WWII perforated matting was the same type used in the project, but not the planes.

1LT Ament was in charge of the project and did an excellent job of keeping the operation together. One time, I was out of radio range to and from the Battalion headquarters sixty miles and a mountain range away. So, he made a very difficult dicision that kept the project going and on time.

At one point after I had returned, the soldiers became fed-up and refused to work. I gathered everyone together and listened to their complaints. The heat was unbearable. The steel matting was hot to handle. They were thirsty and hungry. The sledgehammer handles were breaking (supplying wooden handles was a logistical nightmare. Who in the supply chain could believe one company needed so many handles?)

I told them that WE, all of us, must complete the work on time. We can work at night if needed to reduce the heat problem. Or we can work in the early morning, take a break, and then work the late afternoon into the evening. Or some other combination. Not finishing the project was not an option. After much complaining and discussion about lights at night with generators going, they as a group suddenly decided that they would keep working during the day.

---

[1] https://upload.wikimedia.org/wikipedia/commons/2/22/404th_Fighter_Squadron_-_P-47_Thunderbolt.jpg

I agreed to get them more water and food. The water was the easy part. But I had no idea how to get more rations. The mess sergeant suggested that there was something in the regulations that might help. Armed with the right regulation, paragraph, and line numbers, I met the Division G-4.(Chief of Division Logistics) At first, he was astounded that I was asking for more rations. He said that his soldiers do harder work and do not get that kind of ration support. I explained to him how hard it was to work non-stop with the sledgehammers in the hot sun and also manually handling the hot steel matting with gloves. Then showed him the regulation authorizing extra rations for the conditions my company endured. Reluctant and yet relieved he had a basis to make the decision, so he agreed.

The mess sergeant quickly obtained sufficient food for the effort. Additionally, being a resourceful person he baked snacks and the soldiers rapidly worked to finish the project.

However, the NCOs were upset with me. They were emphatic that I should not have allowed the soldiers to refuse to work. That I should have told them to quit complaining, get back to work, and do their jobs. This is not civilian life. This is the Army, and they vigorously told me "You are in charge, not them!"

I responded that, "Yes, I could have done that. Then I would be pushing you guys to push your soldiers to keep working. That would have gone on for days and days. Now, they are all willingly working. You do not have to push them, and I do not have to push you to push them all day long for at least another week."

They didn't like that answer; yet, they reluctantly accepted it as compared to the alternative which would have involved me beating on them and they would have been forced to push hungry, thirsty troops to work hard in the hot sun.

Drugs

Drugs were a major problem. My crude estimate upon assumption of command was that fifty percent or more of the company was probably using heroin. Every morning, I would walk around the compound counting the small thumb size empty vials. Taking that count and dividing by 180 soldiers gave an estimate of 50 to 80% at times. By the time I left command it was down to 30 percent. It would be nice to think that my actions contributed to the decline. Perhaps it did; perhaps not. Most likely the decline happened because the soldiers rotated out of the unit and went back to the US.

About once a week someone would overdose. Medivac transported the soldiers to a hospital for detox. Then a week or two later they would return haggard, worn out, thin, and with greyish looking skin. Frequently at morning formations, I would give a short talk about the damage drugs did to the body.

I also informally chatted with small groups of soldiers on various project sites mentioning the issues with using drugs, sex with prostitutes, and listening to their concerns about why the US was in Vietnam, when the news media and letters from home were probably saying that the country was not supporting the war.

Not knowing what else to do to counteract the drug usage, one morning I pulled one of the soldiers who had just returned from drug rehabilitation out of the morning formation. Put him in front of the entire company. I told everyone to "Look at him. Look at him. He looks like a skeleton. His skin looks sick. He can't stand up straight. Yet, he is going to work with you. You will depend upon him for the work and if there is an attack you will depend upon him for defense. This is dangerous to you, his buddies, and himself. This is what happens when you take drugs. If you don't want to look like him, don't take drugs."

No one liked what I said. Several of the soldiers complained to me that I humiliated the guy and should not have done that.

I replied if you have better suggestions on how to reduce drug usage, help the company perform better, and keep more of us healthy, please let me know.

They responded that they would complain to the Inspector General and the Battalion Commander.

I said, it is your right to do so. In the meantime, do let me know how to help you and everyone to get over drug usage. It is not healthy for you personally and it is not safe for the unit.

Some of the NCO's and some of the well-meaning soldiers asked why not court martial them?

My response was that if I court martialed even a quarter of the soldiers for drug usage and then the Army discharged them from the service, there would be more people using the drugs with the hope of getting out of Vietnam earlier. However, since the job must get done, we would have fewer soldiers and that would have created more work for everyone else. I wasn't even concerned about how that would look on my efficiency report.

Perhaps my talks and such actions helped some to stay away from the drugs and others to stop. I have no idea. It is one of those unsolved mysteries. Whatever happened to those soldiers? Did they go back to the US and create happy productive lives? Did they go back, wander around from job to job or get unemployment handouts and wither away in some back alley or Salvation Army home? Did they die early because of health complications from drug usage? I'll never know.

On the one hand, I cared for every one of those guys. Yet on the other hand, I was determined that we would complete the projects on time and to standard. It's easy for me to beat myself up over what I could have done differently. But I wasn't going to spend time doing that.

Race Problems

The issue of race relations was a far bigger problem than that of the situation involving drug usage. In one highly emotional incident, an African American soldier assaulted a White soldier with a lead pipe cracking his skull. The reason why has long been forgotten. Because of his injuries the soldier was hospitalized in Japan. As a result, a some of the White guys wanted to 'Get the n----s.' The African American soldiers were quiet and getting ready for something. Later that night, everything calmed down.

Or rather, everyone was just too tired to care anymore. I settled down to a peaceful sleep next to my field phone. It was clumsy and about the size of a quart milk container. To make a call one turned the crank. It sent an electrical signal through a landline to the switch board operator who would make a connection to whomever I wanted to talk to.

That night the field phone rang, I was nearly asleep in peace. Groggily, I picked up the receiver.

A desperate voice squeaked, "Sir! You got to get down here now!! There is going to be a problem."

I thought, "Shit. Enough. I just want to sleep. When does this end?"

After putting on my boots, I toddled like a drunk to the HQ. On the way I wished that I had had a few drinks. I wearily lumbered into the Orderly Room. There was one light on. I thought, "Maybe that light shouldn't be on. Or is it too bright? I am just too tired to care right now."

The clerk said, "Sorry, Sir. I thought something was going to happen. It is all quiet now. It seemed like there would be trouble. Really, Sir, I am sorry."

"That is ok. You did the right thing. Better to do that than not call and then there is trouble. You did the right thing. Thank you."

With a profound sense of relief, I walked outside into the mild pleasant air. What a relief. I looked up at the black sky filled with jewels. In a moment of reverie, I pondered, "Are there creatures out there fighting a war like this? Might they do something stupid?"

In that instant, in less time than it takes an electron to whiz around a nucleus, ten or fifteen angry White southern soldiers came charging out of the darkness like a defensive football team rushing the quarterback. I thought, "Ah shit. Now what?"

They had clubs, sticks, and perhaps iron bars. Their faces flushed red with anger and drinks. They yelled, "We are going to get those god damn fucking n…. s."

I yelled, "Stop!"

They screamed at me, "We are going to get those fucking n s…." I don't remember how many times they say that. They yelled other words that I have long since forgotten.

Then I sensed something behind me. Keeping an eye on the White soldiers, I turned my head, with almost the same motion of an owl turning its head completely backwards, to see what was behind me. In the thick coal-like darkness something was there. Vaguely, I perceived ten or fifteen African American soldiers who blended into the darkness. They were ready to do Taekwondo which is a Korean form of martial arts using just one's hands, legs, and body. They were ready to fight the tanked-up southern boys. They were quiet. They were ready and sober. They were intent and focused.

Suddenly, I felt as though I had split into three parts. One part of me was hovering about the situation, calmly, coolly seeing the whole scene, describing what was going on. There were drunk guys in front of me with sticks and clubs. There were African American guys behind me ready to do Taekwondo. I was in the middle of them arguing with the drunks while thinking, "How does one argue with drunks? I am glad that they don't have their rifles."

The second part of me continued arguing with the intoxicated soldiers.

The third part of me felt like there was a little Leprechaun sitting on my left shoulder giggling.

The White drunks yelled, "Get out of our way. You don't get out of our way; we will kill you."

The little imp continued giggling and said to me, "Act like Barney Fife when the two big goons were going to beat him up." This refers to an episode from the old tv series, "The Andy Griffith Show," where two big goons got out of the car and threatened Barney. He told them that he stands for more than himself. "If you hurt me, the law will find you and you will be in big trouble."

So, taking the Imp's advice I paraphrased Barney Fife's words, but without the squeaky shaky voice and funny antics. I said to the drunks, "If you kill me, you know what's going to happen. The entire world will collapse on you. One of you will rat on the rest of you and you will be in jail for the rest of your life. If you hurt me, I know EVERY ONE of your names. The world will collapse on you. *Don't do anything that might be stupid.*"

Then the cool hovering part of me mentioned, "Where the hell is anyone to help me?"

The little green suited leprechaun, had switched to my right shoulder. Giggling, he whispered, "You aren't paid enough for this! Hee. Hee. Hee. You ain't paid enough for this."

Suddenly, two big King Kong like sergeants appeared. The White sergeant came from behind the White guys. Flaying his huge arms and hitting the soldiers. He yelled at them, "You god damn fuckers. Get the fuck out of here! Get out of here."

Chapter 6: The 101st Airmobile Div   85

The African American sergeant came from behind the African American guys. He too was flaying his massive muscular arms yelling, "You goddam fucking n…..s get the hell out of here."

As though the leprechaun waved a magic wand, the drunk White soldiers and the Taekwondo ready African American soldiers vanished. I stood alone in the deafening silence. The stars twinkled above as though nothing had happened. A soft gentle cool breeze flowed. I stood alone feeling no one cared about me. Neverthelss, I felt that I had a job to do and I would do it.

A couple of the White soldiers came back to me. They asked, "Are you going to do anything to us?"

I replied, "Just go to bed. Get out of here."

Then I thought, "I might have done something stupid. We all might have done something stupid."

A few days the Battalion Commander, Lietuanant Colone Remus flew in by helicopter. I showed him the projects. He was satisfied. Then just as he was about to board the helicopter, he turned to me. He said, "You should have told me about the incident."

I responded, "Sir, if I had told you about it, you would have wanted an investigation. That would tied up your time, some of your staff, my soldiers, and my time. I resolved the issue. So, we all were able to continue getting the work done."

Pausing for a moment, he nodded. Then boarded the aircraft. Everyone continued getting the missions accomplished on time and to standard.

Retrofit a Mechanized Infantry Unit!

The Company worked as usual on repairing roads, culverts, airfields, and other engineering projects, when we noticed that tanks, APC's, trucks, helicopters, and planes were headed west. It felt like a swarm of locusts was moving. Not knowing what was happening the Company kept on working and wondered what was going on. Later the news circulated that the invasion of Laos had begun.

At the end of the invasion, higher headquarters directed my unit to retrofit a mechanized infantry unit. So, I had several discussions with the 101st and explained my engineering unit capabilities. That there were tasks that my unit was not capable of providing. Eventually we agreed that Company D would provide mess, clothing, entertainment, and other assistance as needed to the Headquarters of the Mechanized Infantry Battalion.

The infantry soldier uniforms were torn, dirty, and appeared to be almost falling off. Another unit retrofitted the infantry equipment and weapons. The infantry soldiers were exhausted and angry about the way the South Vietnamese fought. Fortunately for me, their officers dealt directly with those issues, and I dealt with the issues between the two units.

Barbershop Quartet

The 101st Airmobile Division provided a variety of evening entertainment support such as movies and live shows. So, someone borrowed movies and a reel-to-reel projector. The NCO's setup an area outside to view the movies shown on the side of a building or a wall. The screen was a ten by fifteen-foot wall painted white or with sheets draped over it. I noticed that movies about love and home made many soldiers sad, grumpy, and some nearly started fighting. They liked comedy and war movies.

From time to time, the 101st organized a large show for several thousand soldiers. Professional entertainers sang a variety of songs and told jokes. The audience was exuberantly delighted, applauding, yelling, and wanting more. One time, one of the women entertainers showed the numerous ways of smoking cigarettes. She took off her clothes and made the cigarette glow brightly from her vagina. The audience applauded, shouted approval, and were to say the least amazed and delighted. Some of the religious soldiers were disgusted and felt that shouldn't be done let alone made into a show. However, I thought, "It is okay to kill and people, but not okay to watch this dancer smoke a cigarette from her vagina?" Since then I have wondered if she got vaginal cancer.

Shortly after the mechanized infantry unit arrived, the 101st informed me that the entertainment that night for the engineers and the mechanized infantry soldiers would be a barbershop quartet. I imagined a group of guys with mustaches, straw hats, long sleeve white shirts singing the standard late 1800's barbershop songs. I thought what horny GI's would be interest in that?

However, when the group showed up in my company area, they were twenty-year-old White American women dressed in short shorts and skimpy tops. My heart sank and my whole body felt, "Ah crap. Who sent this? What were they thinking? This is going to be a problem."

So, I chatted with the leader and asked her to not sing, *Jesus Loves the Little Children,* and *I Want to Go Home.* My concern was that the feelings conveyed to lonely soldiers would make them homesick and maybe cry; and possibly, to avoid being embarrassed, they would fight. Or they would be angry that they were in Vietnam when no one at home cares about them, so they'd get depressed and would fight.

Fortunately, not everyone went to the show. There were about two hundred soldiers at the beginning. Others came later and stood around anxious to see the first White round-eyed girls in half a year or more. They became bored and restless with the old-time singing and paid more attention to shapely legs. Some began to mumble and shout, "Take it off. Take it off."

The quartet then sang what I asked them not to sing, *Jesus Loves the Little Children,* and *I Want to Go Home.* Some soldiers struggled to hold back tears. Others began fights. I became nervous that there would two hundred infantry and engineers brawling and crying. Somehow the NCO's from both units on their own rapidly broke up the fights and dispersed the audience.

The leader of the quartet was furious with me. She said she would tell the General that the audience insulted and poorly treated her women.

I replied, "I asked you to not sing those two songs which made these guys homesick. That the infantry guys had just come back from a tough operation hungry, tired, and with uniforms falling off of them. And that they were seeing young attractive White women up close for the first time in months. Then you had them sing *Jesus Loves the Little Children,* and *I Want to Go Home.* That was too much for them. What did you expect? That they were going to happily and clap?

She charged off in a rage. I never heard anything from the 101st Division. From time to time in the next decades I wondered whatever happened to those lovely women. And I wondered, what was the person thinking to have a scantily clad barbershop quartet singing to soldiers coming back from a difficult operation. It didn't help that they were in Vietnam while the US population back home didn't care for them?

Maintenance

When I first came on Active Duty two years prior to my Vietnam tour as a 2LT, 1LT Bob Holyfield and then CPT Roger T. Heiman, and LTC's Guthrie, Thayer, and Remus, taught – no beat into me – the importance of equipment maintenance. Those mundane boring lessons lasted for the rest of my life. They emphatically stressed that with good maintenance a unit can do everything within its capabilities. Without good maintenance the unit is worthless. There will be failure of mission accomplishment.

Consequently, every morning, everyone – everyone – including the clerks and the Chaplain – would do equipment maintenance. Check the oil, grease, water, tires, rust, logbooks, gauges, chains, belts, hydraulic fluids, fuel level, loose bolts, transmission fluid, seepage, and noises. Get under the vehicle. Open the hood and look at the engine. Walk around the vehicle. Look at the tires, the body, the top, the glass, the seats – everything about the vehicle.

The vehicles were bull dozers, graders, scrapers, bucket loaders, wobbly wheel rollers, steel wheel rollers, water trucks, fuel tanker, jeeps, three quarter ton trucks, two and a half ton trucks, dump trucks, concrete trucks, ten-ton tractors, trailers, generators, a rock crusher, and a variety of communications equipment. Boring tasks. Everyone except the Chief Warrant Officer for Maintenance, hated it. But everyone did it.

To paraphrase a former Post Office motto, "Neither rain, nor sleet, nor hail, nor heat, nor cold, will stop equipment maintenance." Get in the equipment, on the equipment, under the equipment. No excuses. First activity. Every day. Day after day. Day after day. It stayed with me for the rest of my life, including doing physical and financial fitness five days a week.

This was boring work. It was easier to make excuses that the project work demanded rapidly getting the equipment moving instead of wasting twenty minutes doing equipment maintenance. Soldiers would try short cuts or reasons to not be at the maintenance time. This was particularly true with rain, wet ground, and even in the morning heat. The farm boys instinctively did the maintenance. However, as a group the draftees reluctantly performed their duties and usually needed encouragement.

There were always several soldiers that figured out ways to create a maintenance problem to avoid doing much work. Some would 'accidentally' loosen a bolt so that fluid would drip out. Drive a vehicle while riding the clutch and thus burn the clutch out. Claim that something needed work when it really did not. The Warrant Officer with his vast experience had ways of minimizing these recalcitrant soldiers. The Chief, instead of letting them have free time or an easy time would instead have them work with the mechanics to fix the vehicle. That meant getting dirty and greasy just like the mechanics. An added benefit was the operators eventually gained some appreciation for maintenance.

Also, without consistent daily maintenance, the twenty-plus year old-World War II and Korean War vintage equipment would break down on a project site or the road. Which would require pulling a maintenance soldier off a job and a finding a vehicle to retrieve or fix the broken vehicle on-site.

The Chief, (all Warrant Officers were called Chief), was the powerhouse, mentor, disciplinarian, and a crucial part of the whole maintenance and repair program. Every single one that I met in maintenance and personnel was top notch. I listened carefully to their comments, complaints, anger, and advice. Sadly, I don't remember their names because they all went with the name of "Chief." If it were possible to find every Chief Warrant Officer that I talked with, I would thank them for all that they did and especially the mentoring to officers that would listen.

I always sought out their advice. In turn they and the company Chief particularly, didn't hesitate to give me advice, counsel me, and at times become irate with me. Without their work and advice the unit would not have been functional, because a unit is no better than the maintenance of the equipment. When the equipment is supported, the unit can function. When the equipment is not maintained, the unit cannot function.

As with every organization, the job wasn't finished until the paperwork was completed. Few liked to do proper paperwork. Yet, it was vital for the operators to list the various issues with the equipment. The Chief and the mechanics during the routine schedule maintenance would check the logbooks for the operator comments. Sometimes the comments were useful. At other times they weren't.

The demand for spare parts was very high. So, there was always a huge amount of parts on order. However, the parts might take days, weeks, and months to arrive. Which for a construction operation, let alone in war, was an eternity. It was just not acceptable. Thus, if the higher ups complained that my projects were behind schedule, it still wasn't acceptable for Chief and me to say that the parts were on order. Yet, we both knew that without parts, the mission would not be carried out. With my agreement, he scrounged up two to three truckloads of spare parts. I didn't ask about where and how he got those parts. Decades later I am still grateful to him.

The Battalion Maintenance Officer would conduct announced and unannounced maintenance inspections. Since we were sixty miles from the HQ, someone would let me know of the unannounced visit a few hours ahead of time. They probably wanted to make sure that my Warrant Officer and I would be present. Fortunately for us, this allowed the Chief to quickly police up the unauthorized parts and store them in a spare trailer or shipping container. A driver would take the vehicle for a ride. After the Battalion Maintenance Officer left, the vehicle would come back, and Chief went on with normal operations.

However, somehow the S-3, Battalion Operations Officer (a Major), found out about the extra parts. On a visit he went ballistic and heatedly said to me. "Those parts are unauthorized. Your collecting those parts hinders the missions of other units in Vietnam. You are slowing up the whole engineering effort by hoarding those parts. Turn those parts in!"

He refused to agree with any of my statements on how what we were doing helped him get the projects done on time. Instead, he kept looking at the big picture.

So, I too looked at the big picture – my big picture. If I did not have the parts, then I wouldn't be able to get all the jobs done on time. The S-3, the Battalion Commander, and the Group Commander would not tolerate any excuses.

This was the overall constant situation that someone would slap my hand for one thing or another. So in this case, I chose avoiding getting my hand slapped for excessive parts or crucifying me for not getting the jobs done on time. It just wasn't possible to do all the directives according to the letter of the regulations.

So while the Chief was standing beside me, I said to the Major, "Yes Sir! I will take care of it."

I can still see Chief's disgruntled face; yet, he wisely contained himself and said nothing. He was visualizing broken down vehicles. Scavenging broken down vehicles to get parts to keep other vehicles running. Which would leave some vehicles permanently stripped of parts and not operational. He must have pictured that I along with the two LT's and Battalion would be screamng at him for not getting the equipment fixed. It was easy to see the depression and anger he was heading towards. Not only that, but his mechanics would also be in a rage because they had taken great pride in keeping the equipment going.

So, after the S-3 left, I said to Chief, "Keep the parts. Figure out a way to better conceal the situation."

A sense of relief flowed through his face and body. Nothing adverse happened to me. The S3 was happy that he stood up for what was right in the big picture. The Chief was satisfied that he had parts to keep the equipment running. I was happy that someone wouldn't crucify me for not getting the job done and that I wouldn't be whining because of a lack of spare parts. Perhaps someone would blame some unknown officer in Vietnam, or Germany, or the US, or even outer space for not getting the job done due to a lack of spare parts. Perhaps.

The "Notebook"

Once a week the Battalion Commander had his staff and Company Commanders meet. That meant a two-hour jeep ride on a two-lane paved road. The weather, the sites, and the villages were usually pleasant on the way to the meeting. My driver and I would pass through some villages, go up the Hai Van Pass through the mountains, and then down into Danang for the meeting.

Figure 6-3 "The Notebook"

The Battalion Commander met with all the staff, including the special staff, and the Company Commanders. The information was for all of us and in turn we would make a few comments for the benefit of the group. Then he would meet with each of the Commanders individually. We rapidly figured out how many topics he wanted to discuss with us. He used a stenographer's notebook to write down issues that he wanted to discuss with his staff officers and with the various commanders.

He constantly carried a stenographer's notebook that was six by nine inches[2]. A line down the middle divided each page. On the left he would write some topic to discuss with one of his officers. On the right side would be the resolution or follow-up. The topics were listed as the issues occurred. So, one officer would have items scattered among many different pages.

Lieutenant Colonel Remus placed paperclips next to each action item. The staff paperclips were on the left. The Commander paperclips were on the right. The location of the paperclip on the page indicated a topic for that officer. We learned to roughly count the paperclips in our location on the page. Mine was on the lower right. A lot of paper clips showed a long meeting and possibly trouble. It was a very efficient system and one that I used on and off for the next fifty years depending on the type of work that I was doing.

I made a very definite point on every trip I made to the Battalion Headquarters, to visit 1LT Chuck Stewart in the Battalion Operations (S3). The discussions were very useful in helping me clean the cobwebs out of my brain. That is to clarify my thinking. Although we didn't see eye to eye on a variety of issues, I valued his input. Though a Major was the S3, he and I felt his input into the operations of the Battalion was invaluable. Consequently, to this day I seek out people with different opinions to clear out whatever cobwebs are lurcking in my brain.

Hai Van Pass (NVA Division Underground Bunker?)

It was not unusual on the return trip from the Battalion to the Company to have delays. In the Hai Van Pass there would be a party (An ambush. It was called a party because so many people would show up). So, that would block the road further ahead forcing vehicles to stop. Some people would get impatient and try to drive in the other lane where there wasn't

---

[2] https://www.shoplet.com/National-Standard-Spiral-Steno-Book/RED36646/spdv?pt=rk_frg_pla&ppp=g_eYo5C4lP-6LA&gclid=Cj0KCQjwsLWDBhCmARIsAPSL3_0QZf8BagSY0YIqP3cU__8OCZU9zP86_udwuT4J1qV30Od5ZyY-2kgaAvL-NEALw_wcB

Figure 6-4 *Modern view of Hai Van Pass. But no guard rails nor painted road lines. 2-3 hours from Company to Battalion.*

any traffic. I wondered why they were so anxious to get by this part of the traffic jam. The experienced Transportation Corps drivers would prevent that by moving their vehicles to block others from trying to get ahead. That would also keep their convoy integrity and prevent ambush problems later.

Also, those that were trying to get by the stopped vehicles would just get closer to the fighting and they did not look like the type of soldiers looking for a fight. They didn't have their rifles or pistols in a position that gave any sign that they were prepared.

Figure 6-5 Steep pull up afterwards

Meanwhile, my driver and I would watch the show the jets put on. They flew high in the sky. Turned into a near vertical dive straight down towards the party[3]. Then release a bomb or two. Somehow the pilot in the space of what appeared to be an inch from the ground turned the jet into a vertical climb. How he and the plane withstood the G-forces astonished me. How many times can the pilot do that before he is hurt? How many times can the jet make such sharp turns before the steel and other components in the jet fail? Such is the mind of an engineer. Not thinking anything of the people in the life and death fight a half mile or so ahead.

Figure 6-6 Attacking straight down. How did they make the turn?

On one of several occasions, while watching the show, I was getting annoyed about the delay which hindered me from getting more of the thousand tasks waiting for me at the company headquarters done. Thinking

---

[3] Free to use: https://pixabay.com/photos/air-force-jet-fighter-military-438465/ https://pixabay.com/service/terms/

about the multitude of issues, I was barely aware of what was happening thirty feet ahead of me on top of the Pass. Two boys about ten years old stopped at each of the trucks and jeeps. They had conversations with the soldiers inside.

Reaching my jeep, they asked, "Hey GI, Coke fifty cents. You want my sista? She good. I have nice vial (of heroin) for you. Make you feel real good. Five dolla." Then they saw my railroad tracks (Captain insignia on my collar that had two silver bars together which looked like railroad tracks). They made surprised faces and hurried to the next vehicle.

Still thinking about the tasks to be done at the Company, I vaguely considered, "Where and how did those two ten-year-old boys get so much ice and Coca-Cola on top of this Pass?" Tired and feeling more pressure about losing time waiting for the party to be over, the thought quietly slipped away.

Decades later I was watching a TV documentary, *"Full Circle with Michael Palin."* It was a multi-part interesting explanation of his 50,000-mile exploration of the Pacific Coast Rim. He started in Alaska, went down the West coast of the Americas, and then up the East Coast of Asia, finishing ten months later back where he started in Alaska. I became extremely interested when he was on the Hai Van Pass. I thought, that is where I was. He explained that during the Vietnam War inside the Hai Van Pass was a North Vietnamese Division Headquarters with a field hospital.

At that moment I thought, "Perhaps the two boys got the ice chest filled with Coca-Cola from the NVA, made some coins for themselves, and gave intelligence to their bosses. Perhaps if I had reported the odd activity of the boys in the Pass, the NVA Division Headquarters could have been attacked and destroyed. Perhaps the course of the War would have been altered. Perhaps the Americans could have won and South Vietnam would have been a democracy. Perhaps."

Something similar actually happened just before the battle of Midway. The Americans benefited from one of the luckiest strokes in military history. A single Japanese destroyer left a telltale wake as it tried to catch up with the fleet. The American plane traced the wake to the Japanese carrier fleet. If the spotter had ignored seeing the wake the outcome might have been different for the war in the Pacific. Who knows? Sometimes the fate of empires turns on a coin toss.[4]

Those are the little pieces of boring intelligence that the intelligence community works with to make recommendations for operations. So, if I had reported my casual observation to the Battalion S2 or to the 101st G2 would they have investigated it? Or would the piece of information have just been one of many such trivia?

---

[4] https://www.quora.com/Did-the-US-Navy-risk-an-unnecessary-total-confrontation-with-the-Japanese-Imperial-Navy-at-Midway-Losing-4-carriers-as-the-Japanese-did-would-probably-force-the-US-to-negotiate-for-a-peace-agreement-Why-not-wait/answer/Richard-Lobb-1#comments Accessed Dec 1, 2022.

Chapter 7
Other Activities

In addition to focusing on getting the engineering work done, there were a multitude of other activities. Here are some of those stories.

Socials with the 326 Engineer Battalion of 101st

To better know the 326 Engineer Battalion that was organic to the 101st, I would join them from time to time at one of their officer parties. They were a highly enthusiastic mission focused group. They loved to tell about their exploits, failures, and issues. It was a great way to let off steam and get to know each other and to quickly and informally resolve issues.

The Battalion Commander would make a few comments about leadership, progress, things to improve, and then we would have drinks. Some of these guys were professionals at drinking and I was not. So, I avoided as much as possible getting too inebriated. It also made traveling by jeep in the dark from their compound back to my unit interesting. Because I had a driver, nothing happened.

Midnight Requisitions

A combat heavy engineer company has significantly more equipment than an airmobile unit. They, like us, needed parts for their equipment. Of course, the standard legal approach was to fill out forms, turn the requisition into the system, and by some magic the needed parts would appear.

However, that would take a long while. So, the 101st 'visited' my company at odd hours of the night. In the morning, mirrors, wrenches, and small items would be missing. It didn't take long to get the feeling that if it wasn't tied down, the 101st would ... the delicate word is *midnight requisition it...* that is steal it.

To prevent the 'midnight requisitions' or at least reduce it, guards wandered around the equipment chasing guys away. It seemed to work. Or at least the complaining went down.

Stupid Side trip

For one of the trips to the Battalion Commander's weekly meeting two hours away, I left my Company area early. Along the way, my driver noticed an interesting dirt trail off to the right side. It was just wide enough for a jeep to navigate it. He had an inquiring mind and said, "Hey, I wonder what is on that road?"

For reasons I don't remember, it seemed interesting to take the detour. I assumed that the trail probably snaked its way through the jungle and back to the main road. So, without further thought I said, "Yeah. What the heck. Let's take it."

Figure 7-1 A group similar to what my driver and I stumbled upon

We slowed down enjoying the trees, blue skies, and sunshine as the tree limbs overhead became thicker. After some twenty minutes, the rough trail made a ninety degree turn to the left. About a hundred feet in front of us were ten to fifteen guerilla soldiers dressed in tattered black pajamas with AK17 rifles. They looked like the ones in shown in the picture.[1] Though the picture is just an example. The leader was obviously staring at the map, turning it around and around trying to figure out which way the road went and where he was on the map.

My driver said, "Oh shit."

I said, "Keep going."

The soldiers were shocked to see us appear out of nowhere. As the jeep approached, they showed surprise and yet instinctively separated to the left and to the right. Again, I mentioned to the driver, "Just keep going." I

---

[1] This image is available from the Collection Database of the Australian War Memorial under the ID Number: P01934.033./

nonchalantly placed my left arm on his shoulder and waved the "V" peace sign at the soldiers. They could have easily reached out and grabbed us.

As we passed by them, I listened for the crack of rifle fire and the thud thud of bullets impacting the jeep and hopefully not us. My driver reached for the stick shift to put the jeep into higher gear and get out of there fast. With my left arm still on his shoulder I said, "Don't change gears. Don't speed up. Don't slow down. Just keep going."

As the road turned left, I said, "Get the fuck out of here."

We were fortunate. Also, I violated procedures just like the two guys did when I took over the company when one was killed, and one was wounded. I should have known better than to have so flippantly agreed to a spontaneous trip. We were lucky while the other two soldiers were not. Such is war. Fractions of an inch or seconds sometimes. I got away with a stupid decision and they didn't. How many times were such events repeated in Vietnam and all the wars in civilization? Probably millions of times. Most turned out all right; but, a few paid some heavy consequences.

Similarly, in civilian life, people make hundreds of spontaneous decisions that are delightful and without any adverse consequences. Yet, there are those few off the cuff actions that result in harm or death. Such is life and is part of the meaning of life.

Court Martial

A 1LT was from the deep South. He was standing in my Company Headquarters.[2] A tall strong African American enlisted soldier, also from the deep South, walked in. The two looked at each other. They were like cats that upon seeing each other their fur would stand up and be ready for a fight. The Lieutenant clenched his fists and kept them at his side. The African American soldier slugged the Lieutenant very hard, knocking him to the floor. From what I heard, it was a very good hit.

I was not in the building at that time. Upon learning of the incident, I decided that striking an officer is a punishment that deserves more disciplinary

---

[2] Normally called the Orderly Room. Probably because everything that goes on there keeps the Company orderly? I just made that up.

action than I was legally allowed to administer. So, I sent the Charge Sheet to the Battalion Commander.

He decided it warranted a Special Court Martial with a military judge who was a lawyer under the provision of Article 90 of the UCMJ (Uniform Code of Military Justice). If convicted the enlisted man could get a dishonorable discharge, dismissal, confinement for more than 1 year, hard labor without confinement for more than 3 months, forfeiture of pay exceeding two-thirds pay per month, or any forfeiture of pay for no more than 1 year.[3]

After the trial, the Battalion Commander called me on the secure radio, informing me that the Judge did not convict the enlisted soldier. She felt that due to the African American soldier's background, when he saw the Lieutenant clinch his fists, he felt that he was going to be hit. So, in self-defense he struck first.

I was furious. But, containing myself said nothing. It was late at night, and I went to bed thinking about how to handle this. Would this be open season on hitting officers? Would there be more racial tensions? Or would this blow over? The result was another brick in the building of increasing racial tensions.

The Battalion Commander was concerned that if the Lieutenant stayed in the unit that he would become a target for more violence. So, he said he was transferring him out of the unit. I was reluctant to lose him because he was an excellent officer who was essential to the success of the company. Also, losing him meant that since 1LT Don Schlotz went home, that I would have no lieutenants in the unit when I was supposed to have four. I reluctantly agreed and thought, "I will figure this out somehow."

Since then, I have regretted not forcefully standing up for him and arguing with the Battalion Commander instead of agreeing with him, even though I could see his point. The Lieutenant could have been attacked, injured or even killed.

Now, over fifty years later, both the judge's decision and my non-support of my lieutenant still bothers me. Sadly, it is one of those things to let go of and move on. Nothing I can do about it. No sense in dwelling in the past.

---

[3] https://www.military.com/benefits/military-legal-matters/courts-martial-explained.html

Glad It Wasn't Me

About two months after I assumed command of the company, several of us were at the usual Battalion Commander's meeting and then having lunch. Someone joined us and said, "The Company B Commander, CPT Ralph Cordon, and several others were killed. They were standing together discussing the road project. A landmine had been buried in the road. When they were near it, someone detonated the mine. It instantly killed five soldiers. Half of Ralph's body was blown fifty feet down the road."

My immediate thought was, "Glad it wasn't me." My second thought was, "Strange how selfish one's thoughts and feelings are."

The medic, Sergeant Ed MacNeil III, that I previously mentioned when the ammunition dump blew up and had attended to me for very minor cuts, was also killed. Additionally, SP4 Jeffrey Goodrich, and SP5 Joe Larson, SP5 Christoher Neal were killed at the same time.[4] (SP means Specialist rank.)

Again, twists of fate. It could have been me, but it was someone else. How many times in Vietnam and all the wars of civilization does luck save some while others are killed, injured, or permanently crippled?

My Silly Intense Dislike

For the most part, I have always tried to work with everyone that comes along no matter their personality, background, or skills. People are who they are. Some people say that is a fault of mine; while others politely wonder if I shouldn't be more forceful with some annoying people.

Uncharacteristic of me, there was one guy I came to despise. He came into the Battalion as a Captain. My first meeting with him I had a neutral feeling. Within a few days, I began to dislike him. After several weeks, my dislike turned to hate. However, being a Christian, I talked myself into just disliking and tolerating him.

When he came into the Battalion, he told the Battalion Commander that he wasn't leaving the Battalion Headquarters compound because he was

---

[4] https://www.vvmf.org/Wall-of-Faces/search/results/?casualtyDateMonth=05&casualtyDateDay=02&casualtyDateYear=1971

too scared to do so, didn't agree with the war, and the Commander could court martial him.

The Commander decided that would be the easy way out. But since the captain was an English major he would be perfect for a desk job handling all the records and correspondence for the Battalion. That way the recalcitrant Captain would get what he wanted, and the Commander would have perfect papers leaving the unit. That was a clever move by my Commander and one that I used many times to resolve conflicts.

However, one day I just couldn't resist my sudden silly intense dislike of him. Battalion headquarters had let me know the Captain would be coming to my Company to inspect the paperwork and to make it better. Since he refused to make the sixty-mile trip by jeep, he managed to get a helicopter ride to my Company. Which seemed odd and I wondered how he managed to arrange such a luxury when I always made the trip by jeep.

Nevertheless, I arranged for two jeeps to pick him up at the landing pad which was not in my compound. Both vehicles had M60 machine guns. Instead of going straight to my HQ, they were to take a circuitous route. At a safe deserted area, they were to start firing the two machine guns for a few minutes and then rapidly return to camp. That I figure would scare the stuffing out of him. It was really a stupid idea; but maybe you can sense my feelings for him.

But…unbeknownst to me the Battalion Commander was also on the chopper. Fortunately, the NCO in charge of picking up the captain wisely decided to nix the escapade. Which on the one hand was wise; though on the other hand, I wasn't able to have the Captain rearrange his pants. Such is life. Strange how feelings and wars are. Why spend such effort on such silly and perhaps dangerous causes? Eventually, I learned to skip such stupid emotional actions and get on with life.

378 Letters

In those days communication with my wife was by an ancient method of letter writing. From the time I left to go to Vietnam until I was back with Marlene, we wrote each other a letter a day. That was a total of 378 letters from each of us for a total of 756 letters. My letters were postage free. Nearly half a century later, we still have them. They are somewhere in some dusty box. Perhaps the mice or tiny bugs have read them as they made homes or digested the prose. Some day Marlene and I might get around to reading them. Nothing much in them. Just love notes to each other and chatter about day-to-day things long since forgotten.

It would take two to three weeks for a letter to go from me to her in Germany where she was teaching the children of American servicemen. She would read the letter and a few days later or perhaps right away write a response. Then two to three weeks later the letter arrived. Sometimes there would be days when no letters arrived and then suddenly a batch of four or five arrived. Marlene might be responding to something that I had asked four weeks earlier. She had my letter in front of her, so she would write a cryptic response. Thus, four to six weeks after I wrote her a letter, I received a response. So, sometimes I couldn't remember what my question or statement was.

Nevertheless, we managed to agree on when and where to meet for a one week vacation in a huge city like Bangkok. Somehow the benevolent Army made all the arrangements.

Rest and Relaxation (R&R)

The Army allowed seven days and six nights of Rest and Relaxation (R&R) at a choice of Hawaii, Australia, Bangkok, and perhaps some other places I forgot about. The flight was free; with the soldiers paying the remaining expenses. Soldiers could take the R&R anytime between the third and ninth month of the tour. Experienced soldiers suggested taking the R&R around the eighth month. Their reasoning was that if one took R&R at the fourth month then when R&R was over, the remaining eight months would feel like a prison sentence. However, if one took R&R in the eighth month, there would only be three months until it was time to go home. Anyone could do that standing on one's head because the time was so short.

Since Marlene was in Germany, choosing Bangkok as a place to meet was convenient, a place we had not been to, and during her vacation on my seventh month. So, we easily agreed on Bangkok. I have no idea how many letters that took to make that agreement. If I was curious enough, I could open up the dusty box buried in the closet.

All I remember is getting to the airport in Vietnam to catch the flight to Bangkok. There was some sort of briefing. The Army always has a briefing. A briefing in the morning, a briefing at night, a briefing before crossing the street. It was no different here. Some guy had a script to read to us. Be careful of crooks and prostitutes; the war might be there too, etc. He mentioned something like, "You might do something stupid. If so, contact blah blah blah." No one cared. We were going to have fun.

At the departure airport a clerk decided to give me a tough time. He didn't like the war, his job, officers, or maybe was having a bad hair day. He gave some reason that I couldn't get on the plane. Instead of dreaming of a delightful time with Marlene, my attention intently focused on the present moment. Marlene was arriving in Bangkok just when my flight from Vietnam would be leaving. I wanted to meet my wife and there was limited time.

If I had to stuff myself in a wheel well of the plane, I was going to be on THAT flight. After much discussion and meeting with those higher up in his food chain, I was on board. Strongly controlling my emotions, I told myself, "Careful, you might do something stupid." Eventually, the clerk handed me the boarding pass with a satisfied look. He'd inconvenienced a captain and gotten away with it. On the other hand, I too was satisfied that I'd managed to get on the flight. Though it did bother me somewhat because that meant some other soldier would be bumped off.

Happy exuberant soldiers were on the flight to Bangkok. It was party time!! The six nights and seven days were pure bliss.

The memory of what we did is vague. Buying a skirt for Marlene and a suit for myself. We toured the city with sanctioned Army tours and ventured out on our own. Saw the reclining Buddha, covered with gold or gold paint. The smiling Buddha laid on its back occupying a city block. The river cruise was delightful with happy Thai people and other foreigners.

It was especially fun to partake in the water festival. The city erupted in celebration of the coming monsoon. Almost all the Thai people poured water on everyone especially foreigners. Children would run after people in the street and douse them with buckets of water. Or they would use toy water pistols. Two Thai children with broad mischievous smiles and buckets of water followed us. They were trying to douse us with water. I kept looking back at them. They laughed. Eventually, they tossed the water on some other American. Adults would use hoses to squirt each other. Bangkok was giggling, happy, and great fun. It was a land of smiles and joy. They seemed like the happiest people in the world, unlike the people in Vietnam.

Marlene mentioned to me that she visited a friend of hers in England. Since the friend knew someone in the American Embassy, Marlene might be able to call me in Vietnam. So, the arrangements were made. The call had to go through multiple connections to Europe, India, Saigon, and to Phu Bai about two miles from where I was stationed. The last operator just couldn't figure out how to make the last bit of the connection. I think that if the connection had been made, I would have broken down and cried.

On the last day in the hotel room, Marlene's silence eerily echoed off the walls. Then she cried.

I asked, "What's wrong?"

More silence. An ominous feeling descended over me like a dentist's lead blanket used on patients when doing X-rays. The weight was unbearable. I asked her to explain. She then said, "I can't tell you."

The sickening feeling for some silly reason morphed into a thought. "She wants a divorce! Now, she tells me on her way to the airport." Over fifty years later, I still have no idea why such a silly idea popped into my head. At times I wonder how random thoughts appear out of no where. Some just as quickly disappear and a few remain in the cobwebs of my brain. Perhaps, this happens to many other people.

In silence we headed to the airport. On the one hand, my heart was heavy with parting from her; on the other hand, I irrationally thought she wanted a divorce. Where that idea came from, I will never know. It had no basis. But I felt it. The flight from Bangkok to Vietnam was heavy with the silence of soldiers. None were hooting, hollering, and getting drinks. None were smiling. We just sat in the seats and stared ahead.

Almost did something stupid

In a few hours, I resumed command of the company. However, I did not focus on my commander responsibilities. I didn't care. I really didn't care about anything. What was the point of doing anything? The one love of my life was leaving and there was nothing I could do. I was in Vietnam and she was on the other side of the continent in Germany. Life was useless and pointless. The only gift I could give her was my life insurance policy. So, I felt with an honorable death I could give her a present of the entire amount of $10,000. Today in 2023 that would be worth about $77,000.

So, as part of my normal work, I would visit various project sites. The areas had to be guarded to prevent or at least alert everyone of a possible attack. The platoon or squad would do work while one or two soldiers sat or walked in the hot sun looking for a possible attack. It was boring work. We 'all knew' that wouldn't happen. At least we believed that. We 'knew' that the North Vietnamese and the Viet Cong 'knew' they were going to win the war; and felt that all the work the US engineers were doing was really for them. Therefore, they would leave us engineers alone. It made sense. It was comforting to think that. But then why do the work? Why not go and directly fight them? It was confusing to think of those strategic questions while handling day-to-day problems and directions.

From time to time one of the sentries would notice something. The procedure was to send a patrol to investigate. But I was convinced that Marlene no longer wanted to be married to me. So when I was on a project site, I'd say to the whomever was in charge, "Keep working. I will check it out myself. No sense in slowing the project down." There was some argument about my going alone. But I hoped by going alone the VC or NVA[5] would kill me.

After a few times of doing this, the First Sergeant said to me, "Sir! If something happens to you, the other men must go out and find you. They too might be hurt or killed. Remember, you have a responsibility to the men.

"Duh. That is right." So, I immediately focused on maintenance, discipline, race relations, drug issues, alcohol issues, and work for the 101st Airmobile Division.

Mess Hall Blownup

Since the Company was co-located with the 101st units, it was a natural target. It was common for three or four rockets to hit the base at various

---

[5] Viet Cong. North Vietnamese Army.

times during the night. Sirens would go off. Soldiers would run to various defensive positions. Nothing would happen and after the all clear signal was sounded we would go back to sleep. These attacks were just harassment so that people would be tired for the next day's work. Creating just enough discomfort so that their enemy would not think clearly and hopefully make some mistakes to hurt or kill Americans. Also, it might lull the Americans thinking that nothing was going to happen and get lazy. Then there would be an attack on an unprepared base.

Since the Company was co-located with the 101st units, it too was affected by these attacks. Nevertheless, I repeatedly told myself and others that compared to the grunts slogging through the jungle, we had it good.

Figure 7-2 Messhall blownup by one rocket. (Courtesy of 1LT Ken Ament)

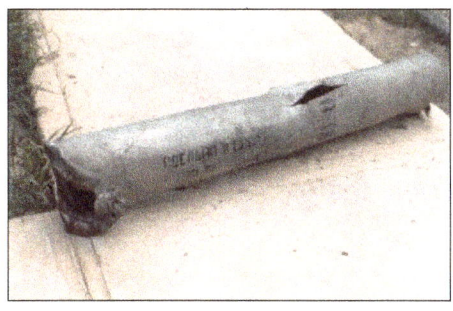

Figure 7-3 Russian Rocket that demolished the mess hall. Courtesy of 1LT Ken Ament.

One time around two or three o'clock in the morning, I heard a rocket exploding and the sirens going off. Then another rocket explosion. I thought, here we go again. It was tempting to sleep in; but one never knew if this would be the beginning of something major or something that would affect the unit. So, I, along with the rest of the unit, put on our gear and began moving. Then the third rocket hit my mess hall.

When the base sounded the all clear, I asked Top, the First Sergeant, to make a headcount verifying that everyone was safe and accounted for. A brief time later he mentioned that only the night baker who worked in the mess hall was missing. So, a group of soldiers carefully picked through the collapsed building in search of him. After a half an hour, someone said, "We found him!"

He was in a bunker and refused to come out. I told Top to, "leave him there and send everyone back to bed. People need their rest, and we'll deal with it in the morning."

In the morning, Top said that the baker refused to come out of the bunker. I said to leave him there thinking that he would come out on his own. However, by the next day, he was still refusing to come out. I chatted with him to find out what was going on, he said to me, "Sir. I have only three days left until I leave to go home. I am too close to be leaving to be killed now."

I told Top, "We can't leave him in the bunker until he goes home. The rest of the guys will begin to wonder, and it isn't good for morale. Three more days creates a problem. Transfer him out now."

Top responded, "Sir we don't have his orders yet."

"You know what the orders look like. Write up the orders, make up some numbers, I will sign it. Get him on a truck and out of here today."

He gave me one of those are you sure looks. I said, "Don't worry about it. No one will figure it out." So, the baker left that day.

I wondered if the system caught him with fake orders at the airport trying to get on a plane home. Never heard anything back, nor did I inquire. Another mystery.

Chapter 8
Going Home

365 Days

Many soldiers carried calendars counting the days until they would go home. Sometimes the calendars were embellished with a black and white drawing of Snoopy as a World War I Flying Ace trying to outfly the Red Baron. Snoopy in the drawing said, "To hell with the Red Baron. I'm going HOME!". Others just mentally counted.

What a way to fight a war.

Nevertheless, we did our jobs well. We basically trusted the US that if were hurt, we would be taken care of, and if we were killed our families too would receive compensation.

Figure 8-1 Many soldiers carried a photo of Snoopy with their count down calendars. It showed their main hope – go home.

Because of copyrights Snoopy is blurred out.

Perhaps

On September 6, 1971, my Company command responsibilities ended with the change of command ceremony which consisted of passing the company flag from me to the First Sergeant, to the new commander, CPT Mike Sells. After a two-hour jeep ride, I was at Battalion headquarters in Da Nang by the South China Sea.

A day or two after arriving I was lightheartedly walking in the compound of the Battalion headquarters. The relief from responsibilities, issues to work on, and the freedom to do nothing was strange. Part of me wanted to do something useful. The Executive Officer asked me to review the Standard Operating Procedures (SOP). It had what, how and when to do various activities and reports. However, part of me was just too mentally and physically tired. All I wanted was to be home with Marlene, to enjoy her company, and to get on with my life. For the first time I noticed and enjoyed the blue sky, the smell of fresh sea air, a gentle breeze caressing my cheeks, the delightful temperature, the shade of the palm trees, and the soothing sound of the ocean swells. Two hundred feet away, the gleaming enticing ocean was beckoning me to swim in the ocean which was off limits.

A soldier new to the Battalion and Vietnam happened to be walking with me. He was a bit nervous about the war and the next assignment. Most likely I reassured him by saying that if he followed the procedures, did his job, and listened to his leaders, the chances are good that he'd go home when his tour of duty was over.

Figure 8-2 Type of Chinook Helicopter that broke apart above me

Then suddenly, an ear-splitting loud thunder bolt directly above me at tree top level snapped my head upwards. A Chinook helicopter with two rotor blades was splitting like an egg into two parts as it continued forward into the South China Sea.

I sharply said to the soldier, "Come on. Let's go." Several more soldiers further ahead of us also ran into the water to help guys out of the chopper.

Moments later, a few disoriented passengers and crew stumbled out. One soldier was face down in the water. I lifted him up. But it was only his top half. One of the rotors had sliced him apart at the waist. I passed the top half to the new guy on my left. I said, "Here. You take him." His face turned paper white. I tiredly thought, "Don't faint. Where is his other half? Welcome to Vietnam."

By then two soldiers had climbed onto the Chopper cockpit windows. They slammed their fists against the glass trying to break the windshield and get the pilots out. We didn't know there was an emergency release lever outside right next to them. Just reach over and pull the lever and the window would have popped out.

A wave lifted me up to the left front window. I grabbed something. Looked into the cockpit. The pilots strapped into their seats were peacefully sleeping as the waves were like a mother gently washing vomit off her babies.

I slid back into the water leaving the guys pounding on the windows with their fists and went around the left side of the chopper towards the jagged opening. I swam and walked inside. The waves repeatedly slammed my head against the roof and then let me down. The roar of the waves muffled the yells of soldiers outside. The closer I got to the pilots, the darker the cabin became when water rushed in. When the water rushed out of the cabin, there was twilight. As the surf relentlessly dragged the chopper out to sea, the water level rose. I had trouble breathing in between the rise and fall of the waves. The cabin visibility increasingly became like the end of evening twilight.

Wires, ropes, and debris tangled my legs. The water washed away my regulation booney hat that I had wanted to take home. Fortunately, the jungle boots and uniform prevented cuts and punctures. I should have pulled the rolled-up sleeves of the jungle fatigues down to prevent arm injury. In the haste I didn't. Instead, I just charged into the main cabin while swallowing saltwater mixed with jet fuel. In the intermittent darkness, the water pushed me up, down, and sideways. I repeatedly untangled myself while swimming, walking, stumbling, swallowing more fuel while tasting saltwater, and banging my head against the ceiling.

I became disoriented, confused, and tired. Just about reaching the cockpit where the pilots were, an involuntary vomiting reflex started. I began hesitating. Without thinking, I turned around, headed for the dim sunlight to where the chopper had split open.

As I neared the jagged opening, a wave as though the ocean was welcoming me to join the pilots lifted me up towards the ragged sheet metal. Frantically, I swam away from metal razor edges, breathing hard, and spitting out fluid but not quite vomiting.

By that time, more soldiers came to help. Some brought a long thick rope to prevent the surf from dragging the chopper further out to sea. Someone

Figure 8-3 Lt William Hatcher with M79 Grenade Launcher and wearing a boonie hat. "Used with permission by LT William Hatcher.

was yelling to get the 50 caliber machine guns and other armaments out of the chopper to prevent the VC from using them against us. For some reason I became focused on not having my bonnie hat. That bothered me. I liked that hat. I wanted that hat. But the waves washed it away.

With all that had happened why did I focus upon that hat? There was no thought about the two guys who perished. No thought about the dozens of other people that helped prevent the waves from dragging the chopper and the two dead pilots out to sea. Soaking wet, dejectedly I walked off the beach, showered, changed my clothes, and forgot about the hat, at least until writing this story.

However, for over five decades I still saw the pilots peacefully strapped into their seats with the ocean caressingly welcoming them. Their skin tone was normal. Not like that of the greyish corpses dumped along a road waiting for retrieval. Those two were the only ones who had died that bothered me. The rest of the dead were like dead rabbits, squirrels, or birds lying along the side of an American road. Don't even notice those nor care. Yet, I still care about my boonie hat and those two pilots.

Sometimes, I wonder if I had pushed harder maybe they could have lived, had families, and enjoyed life. But most likely I would have drowned in the cockpit trying to release a complicated harness I knew nothing about and trying to punch the windows out not knowing that there was an emergency latch inches from their heads. Or perhaps I could have dragged the two pilots through the cabin towards the ripped open chopper, through the entangling wires as more contaminated water filled the cabin and our lungs. Then perhaps others could have lugged the unconscious and probably dead pilots to shore and performed CPR.

Still, I could have done more. Perhaps.

Though not really. Perhaps.

One of those things.

Who cares…?

Vietnam.

So,

we did our jobs,

we went home,

and nobody cared.

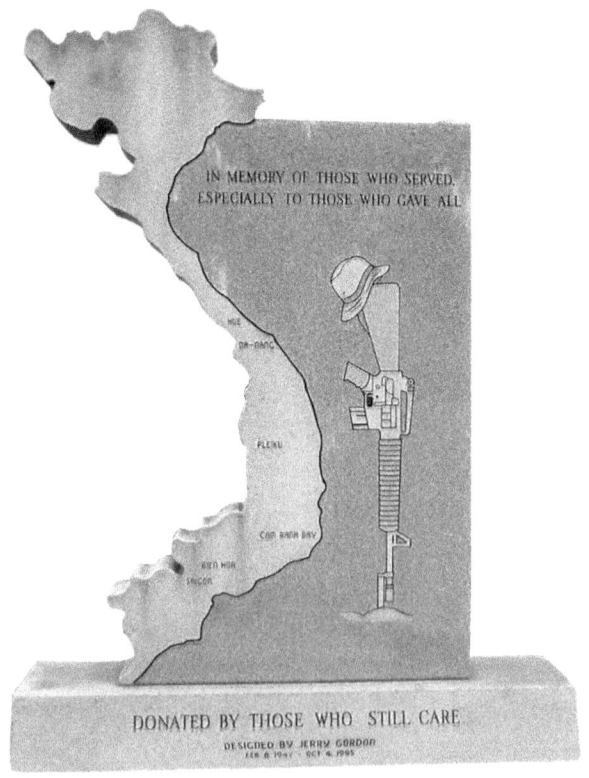

Some of us never really came home.

A Political Suggestion

The fundamental problem of the Vietnam War was the lack of Presidential, Congressional, business, and religious leadership to sell the war to the US and overseas publics. What would have been the sales pitch? Fight to promote democracy instead of communism in Vietnam and prevent the domino theory from happening in all of Southeast Asia? If President Eisenhower had allowed elections in 1954, then Ho Chi Minh would have been president; the country would be communistic as it is now. AND we are happily trading with. So, the entire US leadership said nothing effective to motivate the world public, the US public, and the soldiers.

When contemplating and conducting a war, the US leadership must have the courage to engage the US and overseas publics. If they don't have the courage to do that, then they shouldn't send the military who will risk their lives, of being wounded, and endure family separations.

If the military is willing to risk having their balls shot off, then the Presidential, Congressional, business, and religious leadership of the US should have the balls to sell the war.

President George H.W. Bush did that when selling the First Gulf War (1990-1991).[1] He got the US Congress to vote on the War and then rallied the US population and 35 countries to liberate Kuwait.

It is the job of the President, the Congress, the politicians, business and religious leaders to sell a vision of the US and the world.

---

[1] https://en.wikipedia.org/wiki/Gulf_War

Annex – Vietnam Statistics

US Vietnam War Personnel Statistics

Compiled from three sources in the order that they are shown:

>   https://www.vva310.org/vietnam-war-statistics
>   https://www.americanwarlibrary.com/personnel/vietvet.htm
>   http://www.uswardogs.org/vietnam-statistics/

Personnel

9,087,000 military personnel served on active duty during the Vietnam Era (28 February 1961 - 7 May 1975)

8,744,000 personnel were on active duty during the war (5 August 1964-28 March 1973)

3,403,100 (including 514,300 offshore) personnel served in the SE Asia Theater (Vietnam, Laos, Cambodia, flight crews based in Thailand and sailors in adjacent South China Sea waters).

2,594,000 personnel served within the borders of South Vietnam (I January 1965 - 28 March 1973)

Another 50,000 men served in Vietnam between 1960 and 1964

Of the 2.6 million, between 1 and 1.6 million (40-60%) either fought in combat, provided close combat support or were at least fairly regularly exposed to enemy attack.

7,484 women served in Vietnam, of whom 6,250 or 83.5% were nurses.

Peak troop strength in Vietnam was 543,482, on 30 April 1969.

Approximately 2,594,000 US Servicemen served in country during the Vietnam War.

1,736,000 were US Army

391,000 were US Marines

293,000 were US Airmen

174,000 were US Sailors (this figure includes the US Coast Guard)

Casualties:

Hostile deaths: 47,359

Non-hostile deaths: 10,797

Total: 58,156 (including men formerly classified as MIA and Mayaguez casualties).

Highest state death rate: West Virginia--84.1. (The national average death rate for males in 1970 was 58.9 per 100,000).

WIA: 303,704 - 153,329 required hospitalization, 50,375 who did not.

Severely disabled: 75,000, 23,214 were classified 100% disabled. 5,283 lost limbs, 1,081 sustained multiple amputations.

Amputation or crippling wounds to the lower extremities were 300% higher than in WWII and 70% higher than in Korea. Multiple amputations occurred at the rate of 18.4% compared to 5.7% in WWII.

MIA: 2,338

POW: 766, of whom 114 died in captivity.

Draftees vs. volunteers:

25% (648,500) of total forces in country were draftees. (66% of U.S. armed forces members were drafted during WWII)

Draftees accounted for 30.4% (17,725) of combat deaths in Vietnam.

Reservists KIA: 5,977

National Guard: 6,140 served; 101 died.

Ethnic background:

88.4% of the men who actually served in Vietnam were Caucasian, 10.6% (275,000) were African American, 1.0% belonged to other races

Died in Vietnam

86.3% were Caucasian (including Hispanics)

12.5% (7,241) were African American.

1.2% belonged to other races

170,000 Hispanics served in Vietnam; 3,070 (5.2%) of whom died there.

Helicopter crew deaths accounted for 10% of ALL Vietnam deaths. Helicopter losses during Lam Son 719 (a mere two months) accounted for 10% of all helicopter losses from 1961-1975.

KIA

86.8% of the men who were KIA were Caucasian

12.1% (5,711) were African American; 1.1% belonged to other races.

14.6% (1,530) of non-combat deaths were African American

34% of African Americans who enlisted volunteered for the combat arms.

Overall, African Americans suffered 12.5% of the deaths in Vietnam when the percentage of African Americans of military age was 13.5% of the population.

Socioeconomic status:

76% of the men sent to Vietnam were from lower middle/working class backgrounds

75% had family incomes above the poverty level

23% had fathers with professional, managerial, or technical occupations.

79% of the men who served in 'Nam had a high school education or better.

63% of Korean vets had completed high school upon separation from the service)

Winning & Losing:

82% of veterans who saw heavy combat strongly believe the war was lost because of a lack of political will.

Nearly 75% of the general public (in 1993) agrees with that.

Age & Honorable Service:

The average age of the G.I. in 'Nam was 19 (26 for WWII)

97% of Vietnam era vets were honorably discharged.

Pride in Service:

91% of veterans of actual combat and 90% of those who saw heavy combat are proud to have served their country.

66% of Viet vets say they would serve again, if called upon.

87% of the public now holds Viet vets in high esteem.

Honorable Service

97% of Vietnam-era veterans were honorably discharged.
91% of actual Vietnam War veterans and 90% of those who saw heavy combat are proud to have served their country.
74% say they would serve again, even knowing the outcome.
87% of the public now holds Vietnam veterans in high esteem.

https://www.vva310.org/vietnam-war-statistics

Falsely Claiming to be Vets

The number of Americans falsely claiming to have served in-country is: 13,853,027. By this census, FOUR OUT OF FIVE WHO CLAIM TO BE VIETNAM VETS ARE NOT.

The best estimate that the Department of Defense can conclude is that between 2,709,918 to 3,173,845 GI's served in-country and in-waters Vietnam between 1954 and 1975 (this does not count the handful of Americans who served in Vietnam during WW2). However, veterans groups estimate that today approximately 9 to 12 million Americans fraudulently claim they served in Vietnam.

After extensive research of various mortality indexes and sources The American War Library estimates that approximately one-third of those who did serve in Vietnam (approximately 850,000) are alive today [18 Aug 2007]. (Vietnam veterans are dying at about the same rate as their WW2 fathers.) Since 1988 The War Library has been working to verify the names of all Americans who served in Vietnam, or were authorized the Vietnam War Service Medal.

IN HARM'S WAY. 28 February 2019
https://www.americanwarlibrary.com/personnel/vietvet.htm

Atrocities

Isolated atrocities committed by American Soldiers produced torrents of outrage from anti-war critics and the news media while Communist atrocities were so common that they received hardly any media mention at all.

The United States sought to minimize and prevent attacks on civilians while North Vietnam made attacks on civilians a centerpiece of its strategy.

From 1957 to 1973, the National Liberation Front assassinated 36,725 Vietnamese and abducted another 58,499. The death squads focused on leaders at the village level and on anyone who improved the lives of the peasants such as medical personnel, social workers, and school teachers. – Nixon Presidential Papers.

http://www.uswardogs.org/vietnam-statistics/

Table of Figures

Figure 1-1 Domino Theory – When the first one falls, they all fall. 5

Figure 1-2 Army Huey helicopter spraying Agent Orange over agricultural land 13

Figure 1-3 A person with birth deformities associated with prenatal exposure to Agent Orange. 15

Figure 1-4 Handicapped children, most of them victims of Agent Orange. 16

Figure 1-5 The aftermath of the Mai Lai massacre showing mostly women and children dead on a road, March 16, 1968 16

Figure 1-6 The photo became a symbol of antiwar movement. 17

Figure 1-7 Typical student protesters marching at the University of Wisconsin-Madison during the Vietnam War era (1965). 17

Figure 1-8 King speaking to an anti-Vietnam war rally at the University of Minnesota in St. Paul, April 27, 1967 18

Figure 1-9 Mohammad Ali versus George Foreman, October 30, 1974, "Rumble in the Jungle." 18

Figure 1-10 Symbolic representation of American dead in Vietnam since Jan. 20th 1969 19

Figure 1-11 Guardsmen killed a rioter during the riots at Kent State University May 4, 1970 19

Figure 1-12 Flyer for antiwar march, Washington, D.C., March 1, 1975. "March First Coalition 20

Figure 1-13 Anti-war protest against the Vietnam War in Washington. . 22

Figure 1-14 Flyier for Jane Fonda's Fuck the Army (FTA) rally 23

Figure 1-15 Jane Fond at Press Conference in Holland. 24

Figure 1-17 Jane Fonda urinal target in some American Legion Post restrooms. 25

Figure 1-16 Jane Fonda seated on a North Vietnamese anti-aircraft gun; and earned her the nickname "Hanoi Jane" 25

Figure 1-18 A 17 year old Jesse Washington lynched. Note the crowd of smiling spectators posing for the camera. July 1916. . . . 30

Figure 1-19 Crowd proudly forms for picture of lynching. Note the legs above the crowd. August 3, 1920. 30

Figure 1-20 Sign at bus station 31

Figure 1-21 Book Cover "Manchild in the Promised Land 32

Figure 1-22 Members of the 2nd Inf. Div. north of the Chongchon River. Sfc. Major Cleveland, weapons squad leader, points out a North Korean position to his machine gun crew in 1950. 33

Figure 1-23 March on Washington 1963 where the "I have a Dream Speech" was given. 33

Figure 1-24 President Lyndon B Johnson signs the 1964 Civil Rights Act as Martin Luther King, Jr., and others, look on. July 2, 1964. 34

Figure 1-25 Detroit Riot 1967 and 7,000 arrested 35

Figure 1-26 The Myth of the Addicted Army made the case that drug addiction was not widespread in the US Army. 36

Figure 1-27 Many soldiers carried Snoopy with their count down calendars. It showed their main hope – go home. . . . .

Because of copyrights Snoopy is blurred out. 39

Figure 2-1 An example of a helicopter shooting a minigun 47

Figure 3-1 Engineers built roads, bridges, and airfields in support of these operations 51

Figure 3-2 A typical result of combat operations which the 84th Eng did not directly support 51

Figure 3-3 Seeing the dead along some roads or fields was a frequent occurrence. 52

Figure 3-4 Cam Rahn to Quy Nhon 52

Figure 3-5 Bong Son Bridge under construction by Company B, 84th Eng Bn (Combat Heavy) Courtesy of the 84th Eng Bn Association Vietnam. 55

Figure 5-1 Ngoan-Muc Pass 69

Figure 5-2 Culvert Construction near Song Pha 69

Figure 5-3 1LT Ament leaving Song Pha 70

Figure 5-4 Arrows show the company movement from Song Pha and then by LST to Phu Bai. 71

Figure 5-5 Loading the LST 71

Figure 5-6 On the LST from Phan Rang to Phu Bai 72

Figure 5-7 On the road from the dock to Phu Bai. 72

Figure 5-8 1LT Ken Ament is pondering the current situation while leading the convoy. The ever present children like to get near the vehicles and help themselves to 'souvenirs' 72

Figure 5-9 Combat Heavy Engineer Company Manning Structure . . . 73

Figure 5-10 A civilian example of the Army equipment. 73

Figure 5-11 Ken Ament, Red Dirt, Phu Bai, 1971. 74

Figure 5-12 Trucks lined up to haul dirt for road repair and construction 74

Figure 5-13 Typical asphalt work 74

Figure 5-14 1LT Ken Ament in Hue. 75

Figure 5-15 1LT Don Schlotz in Hue. 75

Figure 6-1 The jeep on the right with the CO's driver, Inge, in front of the headquarters (Courtesy of 1LT Ken Ament) 77

Figure 6-2 *The WWII perforated matting was the same type used in the project, but not the planes.* 79

Figure 6-3 *"The Notebook"* 91

Figure 6-4 *Modern view of Hai Van Pass. But no guard rails nor painted road lines. 2-3 hours from Company to Battalion* 93

Figure 6-5 *Steep pull up afterwards* 93

Figure 6-6 *Attacking straight down. How did they make the turn?* ... 93

Figure 7-1 *A group similar to what my driver and I stumbled upon.* ... 98

Figure 7-2 *Messhall blownup by one rocket. (Courtesy of 1LT Ken Ament)* 107

Figure 7-3 *Russian Rocket that demolished the mess hall. Courtesy of 1LT Ken Ament.* 107

Figure 8-1 *Many soldiers carried a photo of Snoopy with their count down calendars. It showed their main hope – go home. Because of copyrights Snoopy is blurred out.* 109

Figure 8-2 *Type of Chinook Helicopter that broke apart a bove me* 110

Figure 8-3 *Lt William Hatcher with M79 Grenade Launcher and wearing a boonie hat.* 112

Permissions List

Cover photo courtesy of Marlene Beggs

Figure 1-1 Dominio Theory – When the first one falls, they all fall. https://commons.wikimedia.org/wiki/File:Domino_theory.svg

Figure 1-2 Army Huey helicopter spraying Agent Orange over agricultural land https://commons.wikimedia.org/wiki/File:US-Huey-helicopter-spraying-Agent-Orange-in-Vietnam.jpg https://www.vietnam.ttu.edu/virtualarchive/items.php?item=VA042084

Figure 1-3 A person with birth deformities associated with prenatal exposure to Agent Orange. https://commons.wikimedia.org/wiki/File:Agent_Orange_Deformities_%283786919757%29.jpg

Figure 1-4 Handicapped children, most of them victims of Agent Orange https://commons.wikimedia.org/wiki/File:A_vietnamese_Professor_is_pictured_with_a_group_of_handicapped_children.jpg

Figure 1-5 The aftermath of the Mai Lai massacre showing mostly women and children dead on a road, March 16, 1968 Photo taken by United States Army photographer Ronald L. Haeberle

Figure 1-6 The photo became a symbol of antiwar movement. https://en.wikipedia.org/wiki/Buddhist_crisis. https://en.wikipedia.org/wiki/Buddhist_crisis#/media/File:Th%C3%ADch_Qu%E1%BA%A3ng_%C4%90%E1%BB%A9c_self-immolation.jpg.

Figure 1-7 Typical student protesters marching at the University of Wisconsin-Madison during the Vietnam War era (1965). https://commons.wikimedia.org/wiki/File:Student_Vietnam_War_protesters.JPG#/media/File:Student_Vietnam_War_protesters.JPG

Figure 1-8 King speaking to an anti-Vietnam war rally at the University of Minnesota in St. Paul, April 27, 1967 https://upload.wikimedia.org/wikipedia/commons/2/2d/Martin_Luther_King_Jr_St_Paul_Campus_U_MN.jpg

Figure 1-9 Mohammad Ali versus George Foreman, October 30, 1974, "Rumble in the Jungle. https://commons.wikimedia.org/wiki/Category:Foreman_v_Ali,_30_October_1974#/media/File:Ali_castiga_foreman.jpg By Hartmut Schmidt, Heidelberg - Own work, CC BY-SA 4.0, https://commons.wikimedia.org/w/index.php?curid=93086152 "

Figure 1-10 Symbolic representation of American dead in Vietnam since Jan. 20th 1969 . https://en.wikipedia.org/wiki/Buddhist_crisis https://upload.wikimedia.org/wikipedia/commons/0/0d/%22Lie_down_and_be_counted%22_Anti-Vietnam_War_Demonstration.jpg By Hartmut Schmidt, Heidelberg - Own work, CC BY-SA 4.0, https://commons.wikimedia.org/w/index.php?curid=93086152

Figure 1-11 Guardsmen killed a ri.oter during the riots at Kent State University May 4, 1970 Due to copyright possible issues photo not used. The link is at: https://en.wikipedia.org/wiki/Mary_Ann_Vecchio#/media/File:Kent_State_massacre.jpg

Figure 1-12 Flyer for antiwar march, Washington, D.C., March 1, 1975. "March First Coalition Seattle Municipal Archives from Seattle, WA - Seattle Municipal Archives, Public Domain, https://commons.wikimedia.org/w/index.php?curid=35019367 https://commons.wikimedia.org/w/index.php?curid=35019367#/media/File:Anti-war_march_flyer,_1975_(14878021090).jpg

Figure 1-13 Anti-war protest against the Vietnam War in Washington. Photo by Leena A. Krohn. This file is licensed under the Creative Commons Attribution-Share Alike 3.0 Unported license. https://commons.wikimedia.org/wiki/File:Vietnam_War_protest_in_Washington_DC_April_1971.jpg

Figure 1-14 Flyier for Jane Fonda's Fuck the Army (FTA) rally https://upload.wikimedia.org/wikipedia/commons/thumb/3/3b/Jane_Fonda_1975d.jpg/1024px-Jane_Fonda_1975d.jpg

Table of Figures 129

http://peoplesoralhistoryprojectmc.org/historical-photos-of-activism-monterey-county/ *courtesy of Corey Miller. This ticket is from FTA's performance in Monterey, California (from Peoples Oral History Project Monterey County).*

Figure 1-15 *Jane Fond at Press Conference in Holland* https://upload.wikimedia.org/wikipedia/commons/thumb/3/3b/Jane_Fonda_1975d.jpg/1024px-Jane_Fonda_1975d.jpg https://en.wikipedia.org/wiki/Jane_Fonda#Visit_to_Hanoi

Figure 1-16 *Jane Fonda seated on a North Vietnamese anti-aircraft gun; and earned her the nickname "Hanoi Jane"* https://en.wikipedia.org/wiki/Jane_Fonda#Visit_to_Hanoi *This image is used to illustrate the photograph in question in the "Controversial visit to Hanoi" section of the article Jane Fonda. This photograph was shot by a public affairs officer of the Peoples Republic of Vietnam, and released worldwide for distribution.*

1. *There is no free alternative to give the same information.*
2. *The image is a web resolution sized version of the photograph, and does not limit the copyright owners' rights to distribute the original image.*
3. *The image is being used for informational purposes only, and its use is not believed to detract from the original in any way.*
4. *The image is being used for informational purposes only, and its use is not believed to detract from the original in any way.*

Figure 1-17 *Jane Fonda urinal target in some American Legin Post restrooms. Photo by author Ivan Beggs*

Figure 1-18 *A 17 year-old Jessie Washington lynched. Note the crowd of spectators. July 1916.* https://naacp.org/sites/default/files/styles/embed_image_c/public/images/lynchingofjessewashington.webp?itok=V9vZKfBb

Figure 1-19 *Crowd proudly forms for picture of lynching. Note the legs above the crowd. August 3, 1920.* https://lynchinginamerica.eji.org/report/

Figure 1-20 http://loc.gov/pictures/resource/cph.3b22541/ No known restrictions.

Figure 1-21 https://en.wikipedia.org/wiki/Manchild_in_the_Promised_Land#/media/File:ManchildInThePromisedLand.jpg "qualifies as fair use under the copyright law of the United States."

Figure 1-22 Members of the 2nd Inf. Div. north of the Chongchon River. Sfc. Major Cleveland, weapons squad leader, points out Communist-led North Korean position to his machine gun crew in 1950 https://www.nps.gov/articles/000/executive-order-9981.htm.

Figure 1-23 March on Washington 1963 where the "I have a Dream Speech" was given. Wikimedia Commons by the National Archives and Records Administration https://news.harvard.edu/gazette/story/2013/08/the-dream-50-years-later/

Figure 1-24 President Lyndon B Johnson signs the 1964 Civil Rights Act as Martin Luther King, Jr., and others, look on. July 2, 1964. https://www.nps.gov/gwmp/learn/historyculture/images/LBJ-signs-Civil-Rights-Bill-MLK-and-others-stand-behind-him.png

Figure 1-25 Detroit Riot 1967 and 7,000 arrested https://cdn.britannica.com/70/195570-050-B7D927EF/People-Detroit-1967.jpg Accessed Mar 12, 2023

Figure 1-26 The Myth of the Addicted Army made the case that drug addiction was not widespread in the US Army. https://www.amazon.com/Myth-Addicted-Army-Vietnam-Politics/dp/1558497056 https://m.media-amazon.com/images/W/IMAGERENDERING_521856-T2/images/I/614fdtFABjL.jpg

Figure 1-27 Some soldiers carried photos of Snoopy together with calendars showing the days until they get on the Fredom Bird and go home. A returning home soldier gave me the photo to use. To avoid copyright infringement, I blocked out Snoopy.

Figure 2-1 An example of a helicopter shooting a minigun Copyright © International Ammunition Association
Photo obtained by and Forum Maintained by Aaron Newcomer https://forum.cartridgecollectors.org/t/type-of-7-62x51-ammo-used/46593 November 10, 2022 It is a video. Somehow I use one frame from the video.

Table of Figures 131

Figure 3-1 Engineers built roads, bridges, and airfields in support of these operations the Australian War Memorial Photogra- pher: Michael Coleridge. EKN/67/0130/VN https://s3-ap-southeast-2.amazonaws.com/awm-media/collection/EKN/67/0130/VN/screen/6203335.JPG https://www.awm.gov.au/visit/exhibitions/focus/michael-coleridge

Figure 3-2 A typical result of combat operations which the 84th Eng did not directly support Berm and immediate perimeter area of FSB Brown, 13 May 1970 Image by SP4 Peter Nagurny, 40th Public Information Office, US Army. https://signal439.tripod.com/webonmediacontents/LZ%20 Brown%20aftermath_edited-1.jpg?1382739757596 https://signal439.tripod.com/redcatcher199lib/cambodia.html Go about 1/3 down the site for the photo.

Figure 3-3 Seeing the dead along some roads or fields was a frequent occurrence. Berm and immediate perimeter area of FSB Brown, 13 May 1970 Image by SP4 Peter Nagurny, 40th Public Information Office, US Army. https://signal439.tripod.com/webonmediacontents/LZ%20 Brown%20aftermath_edited-1.jpg?1382739757596 https://signal439.tripod.com/redcatcher199lib/cambodia.html Go about 1/3 down the site for the photo.

Figure 3-4 Cam Rhan to Quy Nhon Created by author through Google Maps

Figure 3-5 Bong Son Bridge under construction by Company B, 84th Eng Bn (Combat Heavy) Courtesy of the 84th Eng Bn Association Vietnam. https://www.84thengineers.com/pictures/bridge/bong%20son.htm Nover 26, 2022. https://www.84thengineers.com/pictures/bridge/bridge3%5B1%5D.jpg

Figure 5-1 Ngoan-Muc Pass (Courtesy of 1LT Ken Ament)

Figure 5-2 Culvert Construction near Song Pha (Courtesy of 1LT Ken Ament)

Figure 5-3 *1LT Ament leaving Song Pha (Courtesy of 1LT Ken Ament)*

Figure 5-4 *Arrows show the company movement from Song Pha and then by LST to Phu Bai. Created by the author, Ivan Beggs*

Figure 5-5 *Loading the LST Permission by (Courtesy of 1LT Ken Ament)*

Figure 5-6 *On the LST from Phan Rang to Phu Bai (Courtesy of 1LT Ken Ament)*

Figure 5-7 *On the road from the dock to Phu (Courtesy of 1LT Ken Ament)*

Figure 5-8 *1LT Ken Ament is pondering the current situation while leading the convoy. The ever present children like to get near the vehicles and help themselves to 'souvenirs' (Courtesy of 1LT Ken Ament)*

Figure 5-9 *Combat Heavy Engineer Company Manning Structure US Army Field Manual 5-11, March 2016 http://www.secondcontinentalarmy.com/wp-content/uploads/2016/03/FM-5-11-Engineer-Troop-Organizations-and-Operations.pdf Page A-"B-57*

Figure 5-10 *A civilian example of the Army equipment. By Jaun Rodrequez https://www.google.com/search?client=firefox-b-1-d&q=heavy+earth+moving+equipment+for+construction+images#imgrc=uNzKbl51X3tyvM Earthmoving Heavy Equipment for Construction https://www.liveabout.com/must-have-earth-moving-construction-heavy-equipment-844586*

Figure 5-11 *Ken Ament, Red Dirt, Phu Bai, 1971 (Courtesy of 1LT Ken Ament)*

Figure 5-12 *Trucks lined up to haul dirt for road repair and construction (Courtesy of 1LT Ken Ament)*

Figure 5-13 *Typical asphalt work (Courtesy of 1LT Ken Ament)*

Figure 5-14 *1LT Ken Ament in Hue. (Courtesy of 1LT Ken Ament)*

Figure 5-15 *1LT Don Schlotz in Hue. (Courtesy of 1LT Ken Ament)*

Figure 6-1 *The jeep on the right with the CO's driver, Inge, in front of the headquarters (Courtesy of 1LT Ken Ament)*

Figure 6-2 *The WWII perforated matting was the same type used in the project.https://upload.wikimedia.org/wikipedia/ commons/2/22/404th_Fighter_Squadron_-_P-47_Thunderbolt.jpg*

Figure 6-3 *"The Notebook" " https://www.shoplet.com/ National-Standard-Spiral-Steno-Book/RED36646/ spdv?pt=rk_frg_pla&ppp=g_eYo5C4lP-6LA&gclid=Cj0KC QjwsLWDBhCmARIsAPSL3_0QZf8BagSY0YIqP3cU__8OC ZU9zP86_udwuT4J1qV30Od5ZyY-2kgaAvLNEALw_wcB*

Figure 6-4 *Modern view of Hai Van Pass. But no guard rails nor painted road lines. 2-3 hours from Company to Battalion. Created by Ivan Beggs from Google Maps*

Figure 6-5 *Steep pull up afterwards https://pixabay.com/photos/air-force-jet-fighter-military-438465/ Free to use per: https://pixabay.com/service/terms/*

Figure 6-6 *Attacking straight down. How did they make the turn? https://pixabay.com/photos/air-force-jet-fighter-military-438465/ Free to use per: https://pixabay.com/service/terms/*

Figure 7-1 *A group similar to what my driver and I stumbled upon This photograph is from an album captured by the soldiers of 1 Platoon, A Company, 7th Battalion, The Royal Australian Regiment (7RAR), during Operation Santa Fe at YS561825 on 1967-11-08. The album was used for propaganda purposes, probably by a political officer from D445, the local Viet Cong (VC) Battalion in Phuoc Tuy Province. (Donor Colonel E.J. O'Donnell). Copyright unknown - orphaned work. This image is available from the Collection Database of the Australian War Memorial under the ID Number: P01934.033 Viet_Cong_soldiers_from_D445_Bn_(AWM_P01934033).png https://en.wikipedia.org/wiki/File:Viet_Cong_soldiers_ from_D445_Bn_(AWM_P01934033).png*

Figure 7-2 Messhall blownup by one rocket. (Courtesy of 1LT Ken Ament)115

Figure 7-3 Russian Rocket that demolished the mess hall. Courtesy of 1LT Ken Ament

Figure 8-1 Many soldiers carried photos of Snoopy & the Red Barron along with a calendar marking the days until they went home. Photo by Ivan Beggs To avoid copyright infringement, I blocked out Snoopy.

Figure 8-2 Type of Chinook Helicopter that broke apart above me. Image by Kevin Lyle courtesy from Pixabay. https://pixabay.com/photos/helicopter-royal-air-force-chinook-354699/. I Inverted the picture for the manuscript. Free download per https://pixabay.com/service/terms/

Figure 8-3 Lt William Hatcher with M79 Grenade Launcher and wearing a boonie hat. Courtesy of 1LT William Hatcher

About the Author

Colonel (Ret, USAR) Ivan Beggs has lived, worked, and traveled in the US, Europe, India, China, Vietnam, and South America. He retired from The Timken Company with the position of Program Manager and from the US Army Reserves with the rank of Colonel, two Bronze Stars, and a Legion of Merit.

Colonel Ivan V Beggs
(Retired, USAR)

Education:

- Brooklyn Technical High School
- BS, Mathematics, Worcester Polytechnic Institute
- MA, Theology, Trinity Lutheran Seminary
- MS, Industrial Engineering, Ohio State University
- MBA, Business, Ohio State University
- Graduate US Army War College.
- Married with four children, four grandchildren.
- Lives in Hendersonville, North Carolina.

Other Books by the Author

Quest for a Meaningful Life through
Christianity, Judaism, Islam, Buddhism, & Hinduism

Fourteen Doubts about Five Religions:
Christianity, Judaism, Islam, Buddhism, & Hinduism

The American Mind in the Age of Trump:
Christian Theocracy versus Secularism;
Corporations versus Capitalistic-Socialism

Der amerikanische Geist im Zeitalter von Trump:
Christlicher Gottesstaat gegen Sakularismus;
Raubtier-Kaitalismus gegen sociale Marktwirstschaft

www.ingramcontent.com/pod-product-compliance
Lightning Source LLC
Chambersburg PA
CBHW061737070526
44585CB00024B/2713